Beekeeping

A Primer on Starting & Keeping a Hive

STERLING, the Sterling logo, STERLING INNOVATION, and the Sterling Innovation
logo are registered trademarks of Sterling Publishing Co., Inc.

2 4 6 8 10 9 7 5 3 1

Published by Sterling Publishing Co., Inc.
387 Park Avenue South, New York, NY 10016

© 2010 by Sterling Publishing Co., Inc.

Distributed in Canada by Sterling Publishing
c/o Canadian Manda Group, 165 Dufferin Street
Toronto, Ontario, Canada M6K 3H6
Distributed in the United Kingdom by GMC Distribution Services
Castle Place, 166 High Street, Lewes, East Sussex, England BN7 1XU
Distributed in Australia by Capricorn Link (Australia) Pty. Ltd.
P.O. Box 704, Windsor, NSW 2756, Australia

Design by Yeon J. Kim

Portions of this book are from: *Birds and Bees*, *Sharp Eyes*, *and Other Papers* by
John Burroughs; *Bramble-Bees and Others* and *The Mason-Bees* by J. Henri Fabre;
and *A Description of the Bar-and-Frame-Hive* by W. Augustus Munn.

All equipment illustrations are courtesy of Brushy Mountain Bee Farm. Equipment
shown, as well as additional supplies for beekeeping, can be purchased directly
through their website: www.brushymountainbeefarm.com

Front cover illustrations: bee art © istockphoto.com/quantum orange;
flower art © Dover Publications
All other illustrations © Dover Publications

Sterling ISBN 978-1-4027-7006-7

For information about custom editions, special sales, premium and
corporate purchases, please contact Sterling Special Sales
Department at 800-805-5489 or specialsales@sterlingpublishing.com.

Beekeeping

A Primer on Starting & Keeping a Hive

By

DOMINIQUE DEVITO

STERLING INNOVATION
An imprint of Sterling Publishing Co., Inc.

New York / London
www.sterlingpublishing.com

Contents

About
Honey Bees
&
Beekeeping

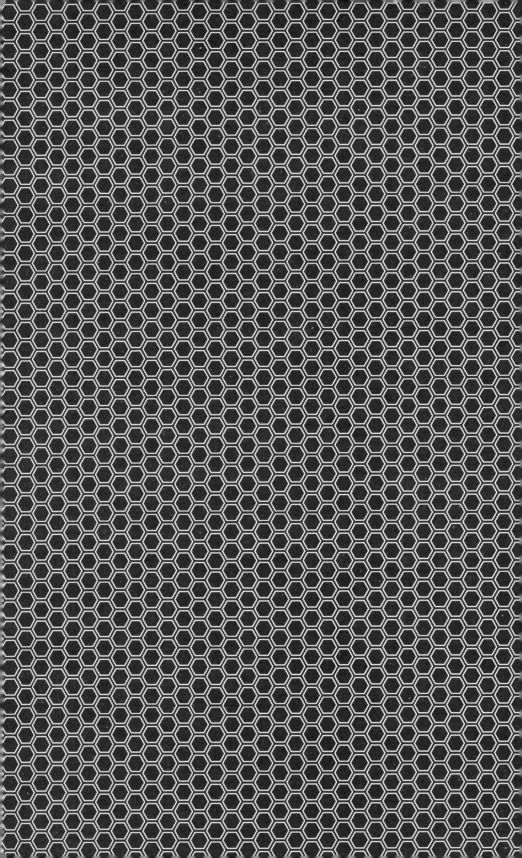

*T*his book is being written at a time when honey bees around the world are getting attention not just for the miraculous and life-sustaining work they do so wonderfully well, but for the very real threat to their existence: colony collapse disorder (CCD). Researchers around the world are trying to determine why whole colonies of bees are abandoning their hives and disappearing without a trace. Besides the personal losses to beekeepers of established hives that are there one day and gone the next, there is the danger that fewer and fewer foods that require pollination will get it—and in the U.S. alone, 80% of pollination is done by honey bees. (More information can be found about CCD in Chapter 4 and in the Beekeeping Resources section.) Essentially, there is no greater time to develop an interest in beekeeping.

Not only will you come to better understand the life of honey bees, but by studying what's happening with your hives, you can be part of the global conversation about the influences that jeopardize the honey bee's very existence.

Every hive counts, and beekeeping can become for you what it is for others who pursue it—a purposeful passion. Before getting started keeping bees, there's lots to know about honey bees themselves.

Scientific Classification of the Honey Bee

The honey bee that will be discussed in this book because it is most frequently kept by beekeepers in the United States is the European honey bee, *Apis mellifera*. Honey bees comprise the genus *Apis* in the family Apidae, order Hymenoptera. They are of the kingdom Animalia, phylum Anthropoda, and class Insecta.

A (BRIEF) HISTORY
of the HONEY BEE

The history of honey bees is as old as that of humankind. They are an ancient insect, for sure. A fossilized piece of pine sap dating 30–40 million years ago contains a bee preserved for the ages and looking remarkably similar to the honey bee we know today. A Spanish cave painting dating to around 6000 BC portrays a man harvesting wild honey as the bees buzz around him.

Honey bees are native to Europe, Asia, and Africa, and most ancient references to honey bees are found in these cultures. The Egyptians, Romans, Greeks, Palestinians, Jews, and many others from these regions celebrated honey and its many blessings. There was the sweetness of honey, which was highly valued, but also the medicinal properties of honey, including its use as an antibacterial healer for the skin. Honey was also fermented to make a sweet wine, or mead, which was drunk at ceremonious occasions (and many others, for sure!).

In whatever ways ancient peoples worshipped and used honey— and developed their beekeeping skills to ensure that their lands were blessed with it—honey and bees were an integral part of their lives. They were so integral, in fact, that when they colonized lands that did not have them, hives were imported. This happened in the Americas,

Australia, and New Zealand. Honey bees came to the United States in the early 1600s, and spread across the country with settlers and others so that they were soon pollinating plants in all of North America. They extended into Central and South America, too. They didn't make it west of the Rocky Mountains by themselves, however—they were brought by ship to California in the 1950s. Honey bees were imported to Australia and New Zealand in the 1800s and were soon an integral part of those countries' growing seasons.

A Prevalent State Insect

So important is the honey bee to so many farmers in the U.S. that over one quarter of our United States have the honey bee as their state insect. They are: Arkansas, Georgia, Kansas, Louisiana, Maine, Mississippi, Missouri, Nebraska, New Jersey, North Carolina, Oklahoma, South Dakota, Tennessee, Utah, Vermont, West Virginia, and Wisconsin.

Fun Facts about Honey Bees

- Bees maintain a temperature of 92–93 degrees Fahrenheit in their central brood nest regardless of the outside temperature.
- Honey bees produce beeswax from eight paired glands on the underside of their abdomen.
- Honey bees must consume about seventeen to twenty pounds of honey to be able to biochemically produce each pound of beeswax.
- Honey bees can fly up to 8.7 miles from their nest in search of food. Usually, however, they fly one or two miles away from their hive to forage on flowers.
- Honey bees are entirely herbivorous when they forage for nectar and pollen but can cannibalize their own brood when stressed.
- Worker honey bees live for about four weeks in the spring or summer but up to six weeks during the winter.

- Honey bees are almost the only bees with hairy compound eyes.
- The queen may lay 600–800 or even 1,500 eggs each day during her three- or four-year lifetime. This daily egg production may equal her own weight. She is constantly fed and groomed by attendant worker bees.
- A populous colony may contain forty thousand to sixty thousand bees during the late spring or early summer.
- The brain of a worker honey bee is about a cubic millimeter but has the densest neuropile tissue of any animal.
- Honey is 80 percent sugars and 20 percent water.
- Honey has been used for millennia as a topical dressing for wounds, since microbes cannot live in it. It also produces hydrogen peroxide. Honey has even been used to embalm bodies such as that of Alexander the Great.
- Fermented honey, known as mead, is the most ancient fermented beverage. The term honeymoon originated with the Norse practice of consuming large quantities of mead during the first month of a marriage.

- Honey bees fly at 15 miles per hour.
- The queen may mate with up to seventeen drones over a one-to-two–day period of mating flights.
- The queen stores the sperm from these matings in her spermatheca, a storage sac; thus she has a lifetime supply and never mates again.
- When the queen bee is about to lay an egg, she can control the flow of sperm to fertilize an egg. Honey bees have an unusual genetic sex determination system known as haplodiploidy. Worker bees are produced from fertilized eggs and have a full (double) set of chromosomes. The males, or drones, develop from unfertilized eggs and are thus haploid with only a single set of chromosomes.

From the Carl Hayden Bee Research Center's "Tribbeal Pursuits," part of the USDA National Agricultural Library

According to the American Garden History blog on beekeeping, "The honey bee was so important in the colonial economy that in 1776, the new state of New Jersey printed its image on its currency. In the 18th and 19th centuries, the beehive became an icon in Freemasonry as a symbol of industry and cooperation. The bee skep is one of the symbols of the state of Utah because it was associated with the honey bee, an early symbol of Mormon pioneer industry and resourcefulness. The beehive is still part of today's Mormon culture."

For centuries, beekeepers kept bees in conical formations made of straw and, sometimes, linen. They were called bee skeps, illustrated above.

Native Americans didn't necessarily take kindly to the invasion of honey bees across the land. The author John Burroughs included this passage in *Birds and Bees*, which appears in the Riverside Literature Series collection of his works:

> The Indian regarded the honey bee as an ill-omen. She was the white man's fly. In fact, she was the epitome of the white man himself. She has the white man's craftiness, his industry, his architectural skill, his neatness and love of system, his foresight, and above all, his eager, miserly habits. The honey bee's great ambition is to be rich, to lay up great stores, to possess the sweet of every flower that blooms. She is more than provident. Enough will not satisfy her, she must have all she can get by hook or by crook.

Burroughs was fascinated by the honey bee:

> There is no creature with which man has surrounded himself that seems so much like a product of civilization, so much like the result of development on special lines and in special fields, as the honey bee. Indeed, a colony of bees, with their neatness and love of order, their division of labor, their public spiritedness, their thrift, their complex economies, and their inordinate love of gain, seems as far removed from a condition of rude nature as does a walled city or a cathedral town.

TWO EARLY AMERICAN PIONEERS

Modern beekeeping had its inception in L. L. Langstroth's development of his movable-frame hive, patented in 1872. Lorenzo Lorraine Langstroth (1810–1895), a native of Philadelphia, at an early age took such an extraordinary interest in insects that he was punished for wearing holes in the knees of his pants while studying ants. He graduated from Yale with distinction in 1831 and was ordained a minister after a course of study at Yale's divinity school. Periods of severe depression limited Langstroth's work as a minister and teacher, but his patient and sensitive observations of the activities in his beloved beehives would change the history of beekeeping.

Langstroth's genius was to recognize the importance of bee space for optimal hive design. Bee space is the term now applied to the roughly $1/4$- to $1/2$-inch corridor that bees require in a hive. When they have less space than that, they plug the gap with propolis, a tough, sticky resinous substance that bees obtain from certain trees; with greater space, they construct a brace comb that connects hive frames. Both of these are great nuisances to beekeepers.

Langstroth's hive, a box enclosing parallel hanging frames, each movable and interchangeable, with all suspended parts being surrounded by bee space, is a perfected home for bees and a perfected

We'll learn more about the basic hive, illustrated above, in Chapter 1.

tool for their keeper. Because of the ease with which bees can be handled with movable frames, Langstroth's hive design has also facilitated the study of bee behavior.

Moses Quinby (1810–1875) is the father of commercial honey production. In his early twenties, Quinby established a beekeeping business, eventually owning 1,200 hives in New York's Mohawk Valley. As a practical man, he sought to make his business more efficient and created several beekeeping aids: one of the first honey extractors in the country, the first useful knife for preparing combs for honey extraction, and the first practical smoker, a hand-operated bellows that blew smoke through a tin firepot. The latter was a very popular improvement over the previous practice of using a smoldering stick to subdue the bees during hive inspection or comb removal. Quinby's original book, *The Mysteries of Beekeeping Explained*, was published in 1853.

Beekeeping Considerations

While you should feel nothing but encouraged to launch into the pastime of beekeeping with gusto, there are some practical aspects of it that you must carefully consider before getting started. These include the amount of time you can devote to beekeeping; the placement and available space for hives; and your budget.

Time

An established hive doesn't actually demand daily attention—though you could certainly give it that. What you will need is the time to start your hobby right, by making critical choices about supplies (see Chapter 1), what your short- and long-term goals are relative to the number of hives you want to have, what you want to do with the honey, whether you want to develop a beeswax-related business, and so on.

While you may be excited to get into a beekeeping hobby, how does your family feel about it? If they're ambivalent, they may soon come around if you help educate them about the proper care and keeping of the bees—and if they can participate in and savor the honey harvest. Don't go into it thinking that this will magically happen, though. They may just as soon become resentful of a new hobby that takes up your leisure time and doesn't include them. In fact, if they're genuinely opposed to the idea from the beginning, reconsider the timing. You may have young children who won't be able to conduct themselves properly around the bees—and who need as much of your time and attention as you can muster. Perhaps the demands of your job(s) take almost all your waking hours, and your partner is pressing you to spend more time with him or her already, or your current lifestyle is already overbooked. On the other hand, it may be the best time, but the important thing is to make beekeeping a conscious choice.

Once you've thought through all these issues (and others discussed below) and you have your hives set up, you will need to check your hives only about once a week. That's assuming that all is going relatively well—the weather is seasonally appropriate, there are no particular disease or pest problems, and so on.

Your Bees Are Your Pets

Just as you are responsible for the humane treatment of a dog, cat, bird, small animal, reptile, fish, or any other kind of living animal you want to keep as a companion in your home or with your family, so you are responsible for the proper—and humane—care of your bees. They are living creatures who depend on you to keep them safe from intruders (bears, for example), keep them healthy, and help maintain the proper life cycle of their hives.

Location

You can keep bees almost anywhere so long as you're sure that their basic needs are met. You don't even have to have a garden of your own to satisfy your bees—they will travel up to a three-mile radius to find what they need. A hive doesn't take up much space, either, and there are many beekeepers living in cities. To optimize life for the bees and to be able to enjoy the hobby, you will need to make sure that:

1. Your hive is fairly accessible to your home. Commercial beekeepers haul hives all over the place, but it's a lot of work, and the honey can be very heavy at harvest time. Your hive(s) should be somewhere where it's convenient for you to check on them fairly frequently, and where you won't have to go too far when it's time to harvest the honey.

2. Your hive has a source of water nearby—but is also in a spot with good drainage! Bees use water to dilute honey when it gets too thick and to cool the hive when it gets too warm, but they do not like to be wet. Consider, too, that if there isn't a nice pond or stream nearby, the bees will get water from a neighbor's garden hose, pool, livestock troughs, and so on—which will be fine for them, but may not make your neighbors too happy. For your bees and your neighbors, if you don't have a water source nearby, you may want to use buckets or a small pool and create and maintain this source.

3. Your hive is in a place that is well ventilated, but not in a place where air gets trapped so that it gets too hot or too cold, or in a place where the wind blows too strongly or gusts. With this in mind, the tops of hills or their valleys are not good locations for beehives.

4. Your hive is in a place with partial sun. Ideally, it should receive dappled sunlight so that there isn't direct sun beating down on it to overheat it or too little sun, which might keep it too cool. A shady spot can keep the hive too damp and the bees will not want to busy themselves.

5. Your hive is facing southeast. This way the bees are waking up with the sun and getting to work.

WHERE ARE *the* FLOWERS?

B ees feed on nectar and pollen, and flowers are their source. If you want to know what plants are contributing to the makeup of your bees' honey, take note of the flowering plants within five miles of you. These will of course vary by region, but prevalent sources across North America include clover, aster, fruit trees, berry bushes, goldenrod, dandelion, ragweed, mustard, cotton, and so on.

Spray Warning

If you live near commercial farms for fruit or cotton crops, consider what pesticides they may be using on them. Talk to the farmers there about their sprays' toxicity to bees; double-check the manufacturer's websites if possible. The last thing you want to do is invest in a hive only to learn that the farmer a few miles down the road sprays something that can kill bees. In Chapter 4, there's a discussion of diseases affecting bees where you can learn more about this issue.

Now you know some interesting things about honey bees and about beekeeping—and there's always more to learn! Finding a beekeeping club in your area will keep you up to date on advances in the hobby and connect you to a like-minded group—not to mention be a great support system for you. To close this chapter, here's another passage from author John Burroughs's essay, *Birds and Bees*, in which he shares his observations of bee behavior:

> Among the humbler plants, let me not forget the dandelion that so early dots the sunny slopes, and upon which the bee languidly grazes, wallowing to his knees in the golden but not over-succulent pasturage. From the blooming rye and wheat the bee gathers pollen, also from the obscure blossoms of Indian corn. Among weeds, catnip is the great favorite. It lasts nearly the whole season and yields richly. It could no doubt be profitably cultivated in some localities, and catnip honey would be a novelty in the market. It would probably partake of the aromatic properties of the plant from which it was derived.

Chapter 1

Supplies You'll Need To Keep Bees

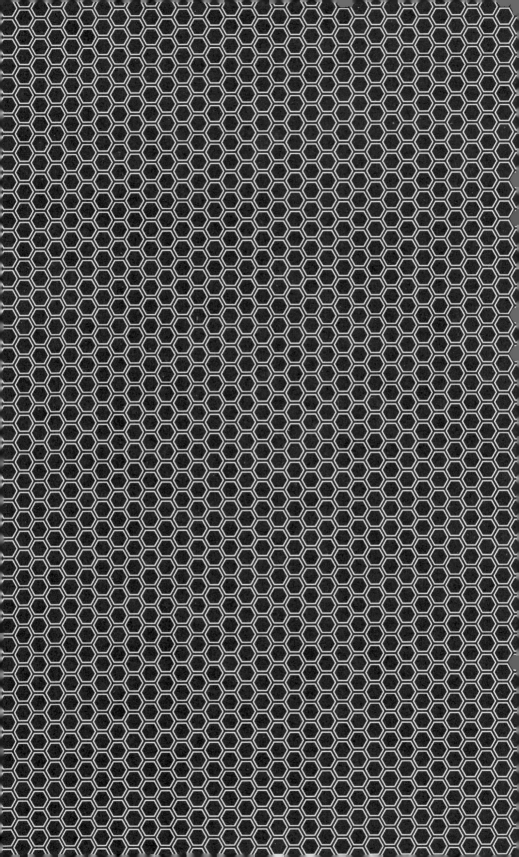

L ike with any hobby or specialty interest, when it comes to supplies, you can really go crazy and load up on all kinds of things. There is certainly room for that in the beekeeping hobby—especially if you want to package your honey in a particular way, or start making candles, or set up multiple hives, or (on the down side) need to deal with a particular disease, which will involve medications. There are many companies that sell beekeeping supplies of all kinds, and you should explore them not only for their range of products and services, but to give you ideas about the kinds of things that are involved in beekeeping. (The names and contact information for several catalog companies are listed in the Resources section.)

In this chapter, we'll stick to the basics: what is essential for getting your beekeeping hobby up and running. This includes:

- Hives and Their Components
- Beekeeper Attire
- Hive Tools
- Bees

The ALL-IMPORTANT HIVE

As you work with the boxes that make up a beehive, be thankful to L. L. Langstroth for inventing them. It wasn't too long ago (1872) that the bee-loving Philadelphian developed and patented this system—and it's been used ever since. Langstroth freed bee-keepers from having to use traditional bee skeps—dome-shaped apparatuses made from thickly braided grass or straw looped together in a dome shape. Gardeners in the eighteenth and nineteenth centuries often included bee skeps in their gardens to ensure that bees were there to pollinate their flowers and produce.

Bee skeps had only one small entryway for the bees, and the only way beekeepers could know what was going on inside the hive was to wait until harvest time. At that point, they would encourage the bees to move to a new beeskep while they dismantled the abandoned one to collect the honey, honeycomb, and wax. This was a messy and wasteful process.

The beauty of the boxed hive system is that it allows beekeepers to monitor the hives. Boxed hives are made up of multiple pieces, each with its own function. The parts of the beehive come together to form a space that is efficient for the bees, and for the beekeeper. There are parts that shouldn't be moved, and parts that can and should be moved. The hive is built from the bottom up, and, when assembled, functions like a luxury apartment building—allowing safe and easy

Outer Cover

Inner Cover

Shallow Super

Medium Super

Queen Excluder

Hive Body
(Brood Chamber)

Bottom Board

Hive Stand

A typical hive setup includes, from top to bottom, (1) outer cover, (2) inner cover, (3) shallow super, (4) medium super, (5) queen excluder, (6) hive body (also called a brood chamber), (7) bottom board, and (8) hive stand.

access and everything an occupant wants and needs (but is easy for the superintendent to maintain). Take a look:

······ OUTER COVER / TELESCOPING TOP ······

The outer cover is a wooden or polystyrene cover that fits on top of the hive. Another option to cover the hive is a telescoping top, which is covered with heavy-duty aluminum, galvanized steel, or plastic to keep rain and snow out of the hive.

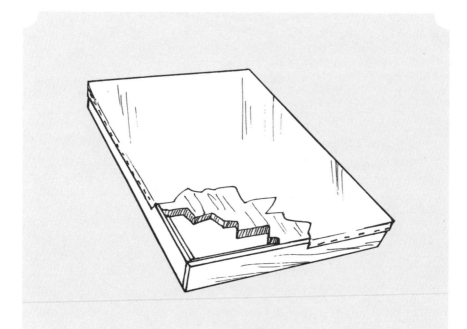

A telescoping top is designed to keep the hive safe and dry. This one shows a reinforced rim, thick plywood, and a securely attached aluminum top.

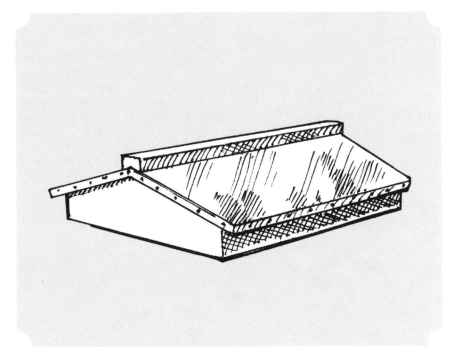

Another option for safely topping a hive is this decorative top from Brushy Mountain Bee Farm. It is designed to give a more finished look to a hive, and is coated with copper, which takes on a beautiful patina over time.

······ INNER COVER ······

The purpose of the inner cover is to provide insulation from heat and cold by covering the super and limiting air movement. There is typically an opening in the inner cover through which you can feed the bees.

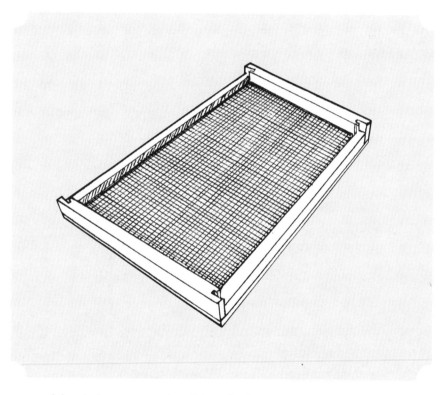

A basic inner cover insulates the hive.

······ **SUPERS** ······

Similar to a hive body but not as deep, the super is where the excess honey that you will be harvesting is made and stored. Honey produced in the hive body needs to be left there for the bees to feed on and live on through the winter. Honey made in the super is yours. For a large, productive hive, a beekeeper may want to add an additional super. These can be stacked. When full, they typically weigh about forty pounds, whereas the hive body can weigh up to eighty or more pounds. The supers are filled with removable frames, as well—anywhere from five or ten per super, depending on the size.

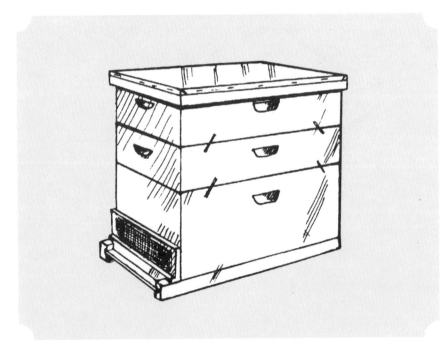

Supers are used to store honey.

······ QUEEN EXCLUDER ······

This is a frame that supports a metal or plastic perforated surface. The holes in the surface are large enough for worker bees to pass through, but not for the queen. The excluder thus prevents the queen from traveling too far up in the hive, limiting her to laying eggs in the hive body. If the queen is able to get into the upper boxes (the "supers"), the worker bees would follow with pollen, which can taint the honey.

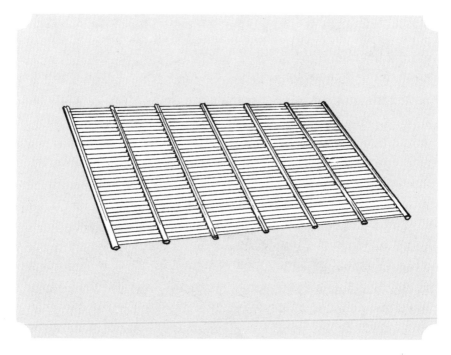

A queen excluder, available in plastic or metal, prevents queens from traveling too far up in the hive.

······ HIVE BODY / BROOD CHAMBER ······

The first hive body box should be a deep one. It typically contains ten removable frames. This box is also called the brood chamber, as it is where the bees live and where the queen lays eggs.

······ BOTTOM BOARD ······

This is a board that serves as the bottom floor of the beehive.

······ HIVE STAND ······

The entire hive sits on a hive stand. This provides a firm base for the hive and elevates it off the ground, making it easier for the bees to get in and out. It is made up of three rails and a landing board, which is the first thing the bees travel across when they return from foraging.

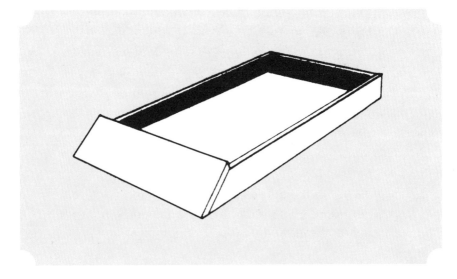

The hive stand supports the entire hive.

Variations of the Basic Hive

There are different-sized supers. It's advisable to start with a large- or medium-sized one and add a small one if necessary. Some beekeepers like to use an escape board, a frame that houses a triangular configuration containing a wire screen. When it comes time to harvest honey, the escape board is placed between the super and the hive body. When the bees are in the hive body, they can't get back into the super with the board in place. This prevents the bees from becoming upset when you remove the honey-laden super.

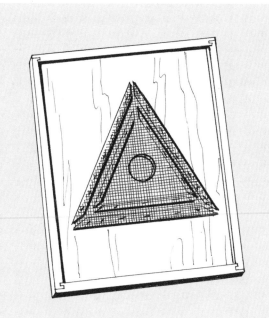

An escape board is placed between the super and hive body during harvesting.

You may want to consider a hive top feeder, too. When there is little or no nectar to forage—especially in the winter—the bees still have to eat. You can feed them through the inner cover, but a hive top feeder makes this job easier. It goes on top of the hive body and under the telescoping top. Its reservoir can hold a gallon or two of sugar syrup.

Materials and Protection

While there are advantages to plastic components, most beekeepers prefer to have wooden hives—and the wood used is typically cypress, as it is highly resistant to rot. If you choose to go with wood, you must paint all the outer surfaces (do not treat any of the interior surfaces of the hive). Choose a light-colored paint made for outdoor surfaces.

Another handy component is an entrance guard or reducer. This is a notched wooden cleat that limits bee access to the hive. It is used to help prevent swarming; it further insulates in colder weather; and it can keep mice out.

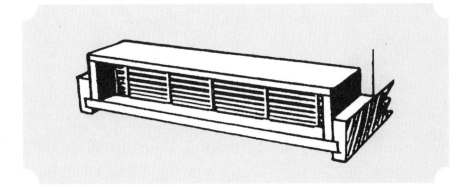

An entrance guard will insulate the hive in cold weather and keep mice out.

Speaking of mice, another pest that afflicts honey bees—with even greater damage—is the varroa mite. There will be a more in-depth discussion of this in Chapter 4, but as part of your hive, you can install a varroa screen. Varroa mites will typically fall off bees as a matter of course, but they will also climb back into the hive and reattach themselves. With a varroa screen in place, the mites that fall off the bees fall through the screen and are unable to get back through it and into the hive or onto the bees. For beekeepers looking to limit the amount of pesticides they use, a varroa screen has been shown to significantly reduce mite populations without pesticides. It is used in place of a bottom board.

OUTFITTING YOURSELF

Honey bees are fairly docile creatures and only sting when they feel threatened or upset. Of course, disturbing the hive is something that will upset them. Though some beekeepers pride themselves on knowing their bees so well they can work on their hives with little protective outerwear, this is certainly not recommended for a beginner beekeeper. These three items should be part of your protective gear whenever you visit your hives:

······ VEIL ······

The purpose of the veil is to protect your head and neck, though some models extend down to the waist. Veils fit over a helmet and include ties that help you secure them around your arms and chest so they stay in place while you work.

······ HELMET ······

Helmets protect the top of your head. Beekeeper helmets are typically ventilated, as much of the hive work is done in warm summer months.

······ GLOVES ······

It makes sense to cover your hands since they are what you will use to work on the hives. There are several varieties of gloves made of different kinds of materials. Ask fellow beekeepers what they prefer; you will probably need to experiment to see which you like best. You may find that the gloves are awkward and clumsy and are more of a hindrance than a help. You may want to reserve the thicker gloves for times when the work is more labor intensive or the bees are more disturbed, such as harvest time or late in the season. For more routine care, though, you may want to use lighter-weight gardening gloves or simply leave the gloves off.

Color Matters

Bees don't like dark colors, so go white or light when outfitting yourself to work around your hives. Most bee-specific protective gear is white, which simplifies things.

Beyond these essentials, there are some basic rules you'll want to follow to further protect yourself. First, wear clothing that covers as much of you as possible. Long pants, long-sleeved shirt, leather or rubber-coated boots. If you wear everyday clothes, secure the ends of your sleeves and bottoms of your pant legs with Velcro straps or elastic that will prevent the bees from accessing your skin. Beekeeping catalogs offer everything from accessories to full-body coverage, and you can choose what you would be most comfortable wearing.

SPECIALTY TOOLS

······ HIVE TOOL ······

This absolutely essential tool is a metal instrument that has a flat side on one end and is bent on the other end. It also has a hole that can be used for pulling out nails or other hardware. It is used to scrape, extract, and manipulate—just about everything. Don't visit your hives without one.

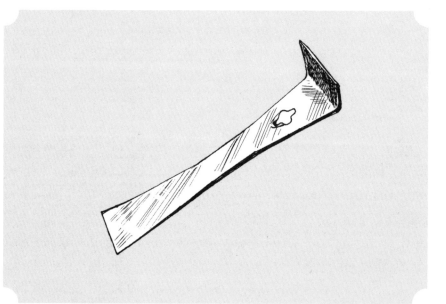

Every beekeeper needs a hive tool, which is used to do everything from separate supers, pull nails, and manipulate hive parts.

······ FRAME GRIP ······

A frame grip is a handy thing, too. It works like a giant ice pick to help you get a good grip on a frame when you're ready to remove it for inspection or collection. There are also frame holders you can attach to the hive body or super to make manipulating the frames easier.

Another tool to make handling the frames in a hive much easier is a frame grip like this one.

······ **SMOKER** ······

Another must-have is a smoker. Even the human portrayed in a cave painting dating to 6000 BC was using smoke to calm the bees so he could work around them. Today's smokers are designed so that the fire chamber is fueled by a simple bellows that releases a cloud of cool smoke. Learn how to work one from an experienced beekeeper before venturing to the hive with one by yourself, and remember that a little smoke goes a long way.

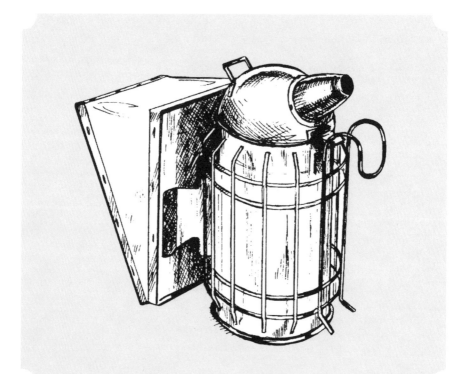

When it comes time to inspect your hives, you'll need a smoker, which safely contains a fire chamber attached to a bellows that allows for the careful application of smoke.

······ **BEE BRUSH** ······

When you remove a frame from the hive, there will inevitably be bees on it. The gentlest way to remove them is with a specially made bee brush. It has super-soft bristles that won't hurt the bees. Our ancestors used goose feathers for this.

The Best *for* Last—Bees!

Yes, your bees! You will need to establish a colony in your hive, and that starts with a queen and her brood. You can order a bee package to get started, which is a queen and about four thousand worker bees per pound. A three- or four-pound package will get a single hive off to a good start. The other way to get a colony is to order a "nuc," or nucleus, which is essentially a complete hive body— frames and all. These are harder to find, as few beekeepers want to part with such an established hive. Besides, it is not difficult to get started with a package, and it's fun to watch your colony develop and grow.

Common honey bees in the Western world are strains of the *Apis millifera*, which evolved from the European or Asian bee, *Apics cerana*. The millifera strains can interbreed, though the *millifera* and *cerana* strains cannot. Which strain you start off with should be determined by the weather extremes in your area (hot and sunny summers or long, cold winters, for example). Today's strains have all been bred to be resistant to certain diseases and to be fairly docile while also being productive. Some are more suited to colder climates and some to warmer climates.

Listed by overall popularity, the four most widely available strains are:

······ ITALIANS ······
(Apis mellifera ligustica)

Since being imported into the U.S. in the late 1800s from the Apennine Peninsula of Italy, these bees have truly "taken off" with beekeepers. Why? Because they are gentle and highly productive. They are fairly disease resistant, and they can survive a cold winter. Their large colonies demand a lot of food in the winter, though. They are yellowish brown with dark bands across their abdomens.

······ CARNOLIANS ······
(Apis mellifera carnica)

Originating in the mountains of Austria and the former Yugoslavia, Carnolians are as gentle as Italians but not quite as productive. On the plus side, the colony size in winter is naturally reduced and is therefore less labor intensive to feed and keep alive. They are equally disease resistant. Carnolians are darker colored than Italians and have gray rather than black bands across their abdomens.

······ CAUCASIANS ······
(*Apis mellifera caucasica*)

Appreciated for their cold-hardiness, the Caucasian bees, while gentle, are not as productive as Italians. They are most prized for their ability to overwinter with minimal upkeep, hailing as they do from the rugged Caucasus Mountains near the Black Sea. Caucasians also tend to be robbers (marauding bees that steal honey from other hives).

······ RUSSIANS ······
(A strain of *Apis mellifera*)

When varroa mites became a real plague to bee colonies across the United States, the search began for a strain that was resistant to this nasty pest. It was discovered that honey bees from the far eastern coast of Russia, near Vladivostok, had been exposed to the mites for more than 150 years. A team of researchers from the USDA Honeybee Research Laboratory in Baton Rouge, Louisiana, identified the strain, conducted breeding experiments, and found that, indeed, it was more resistant to the mite. The strain was imported with much anticipation only recently, and is becoming increasingly popular. Not only is it verroa resistant, but the Russian bee fares well in cold climates.

Bee Hierarchy

Within *Apis mellifera*, there are three distinct castes of bees. They are different in appearance, but more importantly, in function. The three castes are queen, drone, and worker bees. When you order bees, all three castes are included.

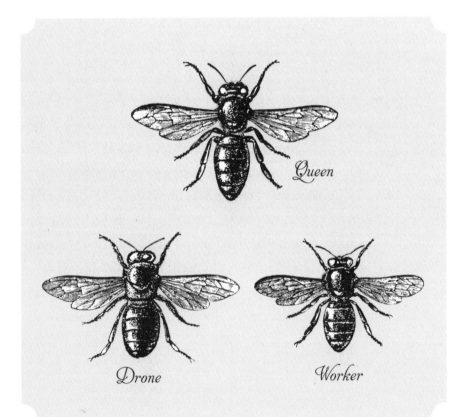

······ THE QUEEN ······

Without a queen there is no hive. She is the heart of the colony, producing the eggs that will hatch new bees, and regulating the colony through scent and behavior. The other bees all take their cues from the queen, and exist to serve her and the hive. Whenever you check your hives, the first thing you'll want to ensure is that your queen is there and looking healthy. When you order bees, the queen is usually marked for easier identification.

······ THE DRONE ······

Drone bees are the least populous of the members of the hive, so you won't (or shouldn't) see too many of them. Smaller than the queen but larger than the worker bees, the drones serve one function only— to mate with the queen. Once that job is complete, the drone dies. Drones have very large eyes and, interestingly, no stinger. As winter approaches, drones are forced from the hive by the worker bees and left to die in the cold.

······ THE WORKER ······

Worker bees are female, but they do not have the fully developed ovaries of a queen and therefore cannot reproduce in the hive. They perform many tasks to grow and protect the hive, not the least of which is foraging and bringing back pollen and nectar. Worker bees have fairly short life spans—six weeks during the most active time of year—and their responsibilities vary during that time from feeding young larvae to supplying water, building comb, warming and protecting eggs, and other functions of the hive.

How a Queen Comes to Be(e)

The worker bees determine which larva of all that are laid and produced will become the next queen of the hive. Once the determination is made, that larva is fed royal jelly. This substance is a potent mix of digested pollen, honey, and nectar along with a special chemical secreted from a gland on the head of each bee. Once she hatches, she takes flight and the drones must fly after and mate with her.

With an understanding of the supplies you'll need, the next thing to do is research supply companies to see which ones you'll want to use. A listing of the major suppliers is in the Resources section, though you should also ask the members of your local beekeeping club for suggestions—or just surf online to see what's available. Have fun!

Chapter 2

Developing
& Sustaining
A Healthy
Hive, By Season

o far you've learned a bit about bees and you've explored the supplies you'll need to keep them. This chapter will explain how to set up your hives and care for your bees throughout the year.

Keeping bees is a bit like growing a garden—you need to get started and set up when it's appropriate. Just as you shouldn't plant corn in midsummer and expect a healthy crop, so you shouldn't try to get a hive started midseason. To get the most enjoyment from your hobby, follow the dictates of climate and the seasons in your area.

EARLY *in the* YEAR

Getting into beekeeping is an excellent New Year's resolution, as January is the time you need to start planning. Almost regardless of where you live in the United States, this time of year is when bees are less active if not inactive. Things will get buzzing from south to north as the days and weeks go by, the days get longer, and the temperatures steadily increase. Local beekeeping clubs suggest the following activities for late winter and early spring:

January/February

Research local beekeeping clubs in your area and contact them about upcoming events and membership. One or more of them should offer an introductory beekeeping class. This is a great place to not only learn more about beekeeping, but also to meet others just like you who are getting started. They can become your friends and allies as you learn the ropes. It's also a place where you'll meet at least a few of the beekeeping club's officers or dedicated members, who will be happy to guide you in whatever ways you need.

You should be deciding and ordering whatever kind of equipment you'll want and need. Read as much as you can about beekeeping to inspire and educate yourself before the bees arrive. Work with the beekeeping club to find and secure the best source for your new colony. Often the clubs bring in starter packs or sometimes know of members who are looking to sell "nucs" (preexisting nucleuses, which are typically already in a hive box).

March/April

Determine where you're going to locate your hive(s) on your property, taking into consideration all the things discussed in Chapter 1. Assemble the hive, and if you are starting with wood, be sure to paint the outside surfaces and the hive stand. Plan to install the bees into the hive around the time that dandelions are blooming in your area.

The Bees' Arrival

If you order bees through the mail, you should alert your local post office as to their arrival date. Chances are the staff have received packages of bees before, but it only helps to alert them and let them know that you will be coming to collect them as soon as they arrive. Ask them, too, to put the box in a cool, dark place until you get there.

Regardless of how long your bees have been traveling, they'll be tired and restless when they arrive. You will hear them buzzing inside the package. You'll be curious to check them out, and you'll likely want to acclimate them to the outdoors as soon as possible. Slow and steady is your best course of action, though. There are some things the bees need before you work with them.

When you leave the post office, try to get home as soon as possible—now is not the time to stop for breakfast with a friend! Get the bees home and put the package where it is cool and dark in your home—a barn, the basement, or the garage. Take off the outside packaging so the bee cage is exposed. It consists of a queen cage on the top and a screened area where the worker bees are. Use a spritzer bottle to gently spray cold water on all the bees. Don't soak them, but spray to cool them and provide them with a drink. Remember, bees are less active when they're cool.

Let the package sit for about an hour to settle the bees. Put a large sheet of plastic under the box, as the next thing you want to spray it with is sugar syrup. This will feed the bees and further relax them. Don't brush the syrup over their cage as you don't want to risk injuring any parts of the bees.

Rehoming Your Bees

Seasoned beekeepers agree that the best time to transfer a package of bees to a new hive is in the late afternoon or early evening. This is the time they'd naturally be seeking to return to the hive and settle in for the night, so it makes sense. If it's an especially hot or windy day, you may want to wait a day. The best conditions are a somewhat cool and still day.

Prepare your hive by removing a few of the frames in the hive box and having everything else ready. Prepare yourself by putting on a long-sleeved shirt, long pants, and veil to cover your face and neck. If you want to cover your hands, use garden gloves so you have some covering but also plenty of dexterity. Fumbling with the bees at this point will only aggravate them.

Take the package of bees to the new hive. Remove the wooden cover and position the queen between two frames in the hive box. Lightly spray the bees with sugar syrup again, then shake and lightly tap the package so the bees settle to the bottom.

Standing beside the hive box, open the package, tilt it up and over the hive box, and gently shake out the bees. They should move toward the queen and into and on the hive box. Let them settle in a bit, then gently replace the frames into the box. Secure the inner cover over the hive box, and slide in the entrance reducer so the opening is minimized. Put the protective outer cover on the hive to keep it dry.

Once you've done all this, your next job is to "let it bee." Give your girls about a week to acclimate. Depending on the weather, they will be eager to get to work. This is a great opportunity to observe them as they settle in. Keeping a distance of at least three or four feet from the hive, watch them go in and out. You may see pollen on some of the workers, and you may even be able to spot a larger drone bee among the colony. Look at, listen to, and even smell what's happening with the bees. It's fascinating!

For the first six weeks or so, continue to feed your bees. Using the feeder tray or some other specialized hive top feeder, continue to provide sugar syrup for the bees. They will not stop foraging because you are providing a food source; rather, the sugar syrup will ensure that they stay healthy and productive while they are acclimating and foraging.

A Basic Recipe for Sugar Syrup

INGREDIENTS
10 cups cold water
5 pounds white granulated sugar

In a large pot, bring the cold water to a rolling boil. Turn off the stove and stir in the sugar. It's a good idea to remove the pot from the stove to prevent any sugar from burning or caramelizing. All you want is for the sugar to dissolve completely. When it is dissolved in the hot water, allow the entire pot to cool to room temperature. Feed the sugar syrup to the hive through the hive top feeder or other system you've installed. Be diligent in cleaning up both in your kitchen and around the hive after making, transporting, and serving the sugar syrup. It is appealing to many other creatures besides the bees! Don't clean up with chemicals. Bring everything back into the house and use soap and hot water to clean everything thoroughly.

This easy-to-use feeding system is a great way to provide sugar syrup or water to the hive. It attaches near the entrance. All you need is a quart-sized glass jar.

SPRING CONSIDERATIONS

May/June

As things start to bloom all around you, the bees will really get into high gear. If you're a morning person, you can see them leaving the hive and heading out into the world, then returning later in the day, often with multicolored pollen.

As their work progresses, you'll want to begin checking the hives on about a weekly basis. You could add a super on top of the hive box for honey production. You should ask fellow beekeepers what they're doing about potential pest problems at this time of year, too. Again, depending on how everything goes, you may even be able to harvest some honey by July, so get ready!

Inspecting the Hives

And now, the moment of truth—your first look inside the hives. Exciting, scary, confusing—it will be all of these things. What it won't be is boring. Consider yourself a beekeeper and savor the experience. Give yourself the advantage of checking the hives on a sunny, warm day. Check on them at a time when you think the bees will be actively working. When they're busy, they're less inclined to pay attention to you.

When you're dressed—but before you put on your helmet, veil, and gloves—get your smoker ready. Think of it as a miniature charcoal grill. You'll want to get a fire going in the bottom with some newspaper and kindling, and once that's going, add larger kindling or make it easy on yourself by using smoker fuel that's available through beekeeping supply companies. They're typically compacted cartridges of cotton fibers or wood nuggets. They're safe to use since they're natural, and not synthetic, which could harm the bees.

Pump the bellows a few times while your smoking material is going, to keep the flames stoked. Put on your protective gear, and walk calmly over to your hive. Gently blow smoke in the entrance and near the hive. Standing to the side of the hive, slowly take off the outer cover, blowing smoke into the hive as you do so. Remove the outer cover, then blow some more smoke down into the hive.

Working slowly and deliberately with your hive tool, continue to pry off the layers of the hive, gently blowing smoke into the hive as you do so. Finally, you'll have the frames exposed and you'll be ready to start your inspection.

Stay Calm!

Sounds simple to someone who isn't standing over a colony of buzzing bees. Beekeepers throughout time have noted that the calmer and more focused you are while tending to your bees, the less disturbed the bees become. A steady and attentive demeanor paired with light-colored clothing and the proper protective gear will keep you safe as you work around your hive.

The frames will be sealed together with propolis, a sticky substance, and you'll have to pry them apart. Start with the second one in from the side you're working from. Lift the frame with both hands and slowly bring it up and out of the hive. Your checklist for inspection is:

1. Is the queen there? Is she alive? Is she healthy?
As the colony grows, finding the queen can become difficult. You'll know everything's okay, though, if you see eggs in the comb. These are fairly easy to identify if you hold the frame up so that light shines through the comb. The eggs look like little grains of rice in the pockets of the comb. If there are eggs, there is a queen.

2. Are there larvae?

When eggs hatch they become larvae on their way to hatching young bees. The larvae look like small white blobs curled up in the bottoms of the comb cells. These are tended to by worker bees designated as nurses who steadily feed the larvae honey and pollen. If you find eggs and larvae, you'll know your queen is doing her thing. If you don't see eggs or larvae, or if the eggs aren't hatching into larvae, there is probably a problem with your queen.

3. Does the comb appear healthy?

The overall health of the comb is indicated by the number of eggs, the compactness of the brood chambers, the color of the wax cap, and a generally healthy appearance. If the comb is coming together in a tightly packed way with a nice pattern of eggs, larvae, and capped brood cells, you'll know your hive is functioning to form.

Recognizing the Queen

Of the incomparable thrill of watching the queen bee intermingle with the workers and drones, the author John Burroughs wrote in *Birds and Bees*:

> The queen, I say, is the mother bee; it is undoubtedly complimenting her to call her a queen and invest her with regal authority, yet she is a superb creature, and looks every inch a queen. It is an event to distinguish her amid the mass of bees when the swarm alights; it awakens a thrill.
>
> Before you have seen a queen you wonder if this or that bee, which seems a little larger than its fellows, is not she, but when you really set eyes upon her you do not doubt for a moment. You know that is the queen. That long, elegant, shining, feminine-looking creature can be none less than royalty. How beautifully her body tapers, how distinguished she looks, how deliberate her movements! The bees do not fall down before her, but caress her and touch her person.

April Showers, May Flowers, Bee-ware!

If you're new to beekeeping, you might be wondering what is especially attractive to your new charges. Learning about the plants they frequent will inform you of the flavors you might find in the honey when you harvest it. Here is a partial list of spring-time bloomers that honey bees are attracted to in the northeastern United States:

Apple trees
Cherry trees
Black locust trees
Buckeye
Chestnut trees
Chives
Dandelion
Hawthorne
Heather
Honeysuckle
Maple
Pear
Phlox
Purple deadnettle
Willow

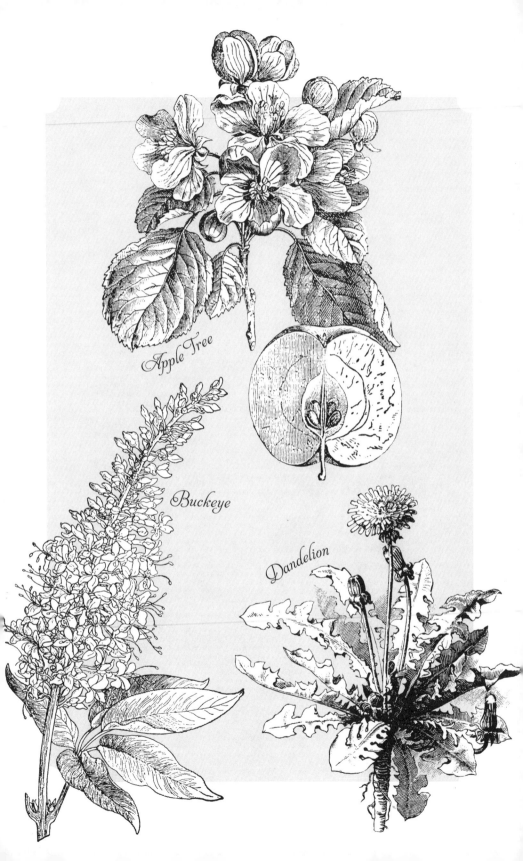

Apple Tree

Buckeye

Dandelion

Pear

Honeysuckle

Chestnut Tree

Maple

Cherry Tree

Willow

Beware a Cold Snap

Depending on where you live, it is very possible that the temperature may be close to 80°F one day in May and then drop below 50°F the next. If that cool, cloudy day is the one you have scheduled to examine your hives, find out what the forecast is for later in the week and hold off until the temperature climbs again and the sun is out. Removing frames on a chilly day exposes the brood to the cold temperatures and can potentially kill them. It will also reduce the temperature of the hive, which imperils the developing larvae.

Hive Alive

It's impossible not to hear the bees working all around you when you go to the hive. Getting a sense for how the bees sound can key you in to how they're doing, too. You will be able to tell whether their humming and buzzing is the normal sound of bees simply working hard, or whether it's a sound that's indicative of something amiss. John Vivian, author of *Keeping Bees*, describes the bees' sound this way:

> You will quickly learn to gauge a colony's mood by its hum. A contented colony makes a sound that varies in

intensity, but remains in a low-frequency range that is pleasant and soothing to human ears, a happy sound appropriate to a warm spring day with fruit trees in bloom. After the opening hum, an unsmoked and happy colony will sound content, and bees that fly up to inspect you will buzz around slowly, acting curious rather than hostile so long as your movements are deliberate.

In advising new hobbyists on how to assess the temper of the bees, he adds,

If several bees come out flying straight and fast to check you, stand still until they depart. Smoke them if they land on you. Alarm a single bee and it can signal the colony that danger's afoot.

Vivian has excellent advice about how the bees' sound can alert you to very real danger, too. He writes,

A hostile colony will warn you in unmistakable terms. The hum becomes loud, shrill and strident, a high-pitched beeeeeeeeeeeeeeeee sound. The vibration rate is unpleasant bordering on fearsome to humans, high enough to cause inner ear discomfort in many animals. It is an adrenaline-generating alarm signal that strikes a primordial chord in humans, the same as a rattlesnake's burrrr or a dog's grrr or an infant's high keening wail. You are best advised to retreat and try again on the next warm and sunny day.

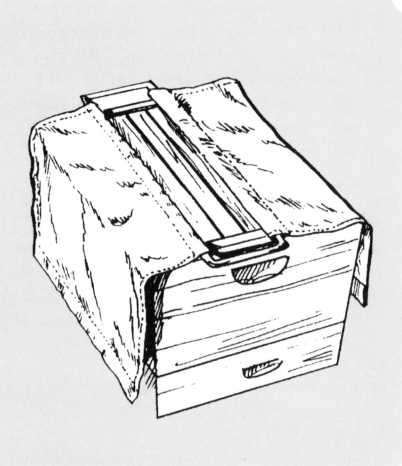

Another way to keep bees quiet while you're working around the hives is with a specially designed manipulation cloth, which has an opening that fits over a single frame while covering the others.

SUMMER CONSIDERATIONS

July/August

If you think that your first colony won't produce enough honey for a harvest, think again! It might not, but you never know, and you should be thinking about how you'll manage this. As summertime progresses and you are checking your frames weekly, you will notice the collection of honey in the comb cells in the upper right- and left-hand corners of the frames toward the center of the hive box. Your busy bees should be extending the comb to six or more of the frames. If things are looking good and the egg and larvae production seems steady and healthy, you should add another hive body box and then a super. Be sure to put the queen excluder between the hive box and the super so that the upper box is only for honey production. Remember, too, that the bees need honey to live on over the winter, so even though you're seeing promising honey production, don't take too much.

While you're inspecting frames at this time of year, you'll need to start monitoring the types of cells that are being nurtured by the nurse bees. Bees are instinctually self-regulating, and if they sense crowding in the colony, they will nurture new queens so that the colony can split. When a colony's numbers swell and a new queen is hatched, the colony will swarm, with a large number of them flying off to be with the new queen.

How Does Honey Happen?

Honey is the liquid that is sealed inside the honeycomb. It is produced from the nectar that bees collect as they forage. When the nectar is brought to the hive, it is about three-quarters water. The bees work to evaporate most of the water, and as they do so, enzymes transform the nectar into honey.

This is rare in first-year beekeeping, but it can happen. Rather than fall victim to a swarm departure, you want to provide your colony with room to grow, which is why you need to monitor the hive for activity. As you observe the cells containing the larvae, you will see that the majority of them are horizontally oriented. The smaller ones house the worker bees, the larger (far fewer) ones house the drones. These cells are vertically oriented and larger than the rest. These are queen cells, also called supersedure cells. If they're located on the top part of the frame, it's an indication that the queen is not performing to peak capacity, so the colony is preparing to replace her. If the supersedure cells are forming on the bottom of the frame, the colony is preparing to swarm.

A Summer Swarm

Swarming is a natural instinct of bees, and beekeepers need to recognize when it's time to do something to prevent swarming. The first thing you want to do is alleviate the overcrowding, and this typically means dividing your hive. Working midday on a warm, sunny day so that most of the worker bees will be off foraging, you'll need to remove several frames with comb and a queen and transfer them to a new hive box. Make sure there's a queen in the original hive box, or install a new or younger queen. If you need a new queen, it's better to introduce one yourself so that the transition is smoother for the colony.

John Burroughs has an insightful reflection on swarming in his essay *Birds and Bees*: "Apparently, every swarm of bees before it leaves the parent hive sends out exploring parties to look up the future home. The woods and groves are searched through and through, and no doubt the privacy of many a squirrel and many a wood mouse is intruded upon. What cozy nooks and retreats they do spy out, so much more attractive than the painted hive in the garden, so much cooler in summer and so much warmer in winter!"

Swarming is most likely to happen through midsummer, so that is when you will need to be diligent about watching for it. Your weekly inspections of the hives should keep you abreast of the situation, and if you're unsure about what you're seeing, consult with another local beekeeper.

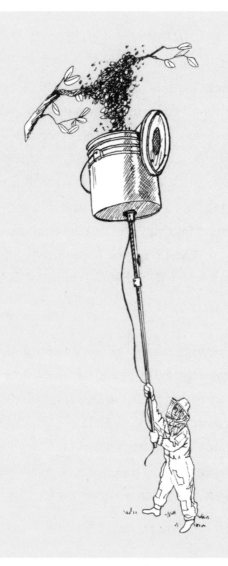

To retrieve a swarm that has settled in the branches of a tree, you can use this contraption designed by beekeeper James Hipps. It's a five-gallon bucket securely attached to a ten-foot conduit with a connector that allows you to close the bucket when the bees have been shaken into it.

Howland Blackiston, author of *Beekeeping for Dummies*, has a handy "7/10 Rule":

- When 7 of the 10 frames in the lower deep are drawn into comb, add a second deep-hive body with frames and foundation.

- When 7 of the 10 frames in the upper deep are drawn into comb, add a queen excluder and a honey super.

- When 7 of the 10 frames in the honey super are drawn into comb, add an additional honey super.

Other summertime concerns include being on the lookout for hive robbers, and harvesting your honey.

Hive Robbers

Bees are always patrolling their environments, and if natural nectar sources become scarce, they will go hunting for any that might be available—including nectar sources in other hives. In the discussion about types of bees earlier in this book, one of the characteristics mentioned was a tendency to rob. It's an aggressive behavior that does come naturally to most bees, so be on the lookout for it.

Identifying a Robbery

If your hive is under attack, the marauders will do everything they can to enter and kill off the bees living in it. Meanwhile, the bees that are there will do everything they can to keep the enemy bees at bay. This results in heavy losses on both sides.

Hives that can't protect themselves risk losing everything—the bees, the honey stores, everything. It is a disaster for the beekeeper (and the hive, of course).

As you get to know your bees, you'll get a feel for their normal behavior and activity. There will be lots of action at the hive when nectar is flowing, and you'll see the bees busily entering and exiting the hive. You will be able to tell the ones that belong there when they return as they will be weighed down with nectar and pollen. The guards at the hive entrance do a good job of patrolling for invaders, too.

When robber bees threaten, everything going on around the hive becomes frenzied. The robber bees themselves may not enter the hive directly, but will fly around the sides looking for opportunities to get past the guards. Guards will be frenetically defending the hive entrance, or may be engaged in battle at the entrance or on the ground in front of the entrance. Robber bees will be leaving the hive with honey stores, so they will be weighted upon exiting the hive.

······ **HANDLING A ROBBERY** ······

The first line of defense against robber bees is to reduce the size of the hive's entrance. This makes it more difficult for the intruder bees to enter and easier for guards to defend. This can be done by putting grass at the entrance to narrow traffic flow. More aggressive actions vary depending on the beekeeper. Here are some suggestions:

- Place a sprinkler on top of the hive. This simulates rainfall and can deter robbers.

- Place a sheet that's been thoroughly soaked in water and then squeezed dry over the entire hive, making sure the sheet reaches to the ground. This prevents robber bees from finding their way in.

- Dilute some Liquid Bee Smoker with water in a spray bottle and spray the bees at the entrance. This usually stops everyone instantly.

······ **PREVENTING A ROBBERY** ······

As with most potentially harmful situations, prevention is the best cure. Guard against a situation where other bees are robbing your hive by maintaining safe and sanitary beekeeping practices. These include diligent handling of any sugar syrup and collected honey, as any open sources will attract bees—yours and others. When working around your hives, be sure that your supers are covered when you remove them, be sure the hole to the inner cover gets closed, and keep the size of the entrance to a minimum.

FALL CONSIDERATIONS

September/October

Fall is an interesting season for beekeepers, because it is a busy one for both keeping up with production of honey and yet slowing down the activity of the hive as winter approaches. There are many late-blooming plants that bees feed on well into September and even early October, depending on your location. For these months, your bees may be as busy as ever and you may even be able to harvest more honey.

Remember that as much as you enjoy the honey—and possibly want to sell it to other honey lovers—the bees must have enough honey in the hive to feed themselves through the winter. If you live in a colder climate, figure on at least sixty pounds; bees in warmer climates may make due with thirty pounds for the winter, but it's better to err on the safe side and leave them plenty. This quantity is typically five to ten full frames of honey.

As the days grow shorter and cooler, the bees will begin to settle down for the winter. When the daytime temperatures fall below about 57°F, the bees will begin to cluster. It is your job to adequately prepare the hive for them so they can survive the winter. As you've been inspecting the hives through the summer, you should have been noticing whether the queen was still actively laying eggs. If you had any kind of indication that your colony might swarm, the bees may have been alerting you to the fact that the queen is not as productive as she should be. For the life of the colony over the winter, it will need a strong queen to survive and to start laying again in early spring.

Some beekeepers will requeen the colony every fall to ensure that there is a young and productive queen in it over the winter. If you have several colonies and only one or two of them appear weak, you may want to combine the colonies. Ask an experienced beekeeper for advice in this area. Before it gets too cold—while temperatures are still in the fifties and sixties—you should feed the bees a two-to-one sugar syrup solution. Start with a small amount and work up, depending on how many bees you have. So for example, into one part boiling water—you may want to start with two cups of water—stir and dissolve two parts sugar, which in this case would be four cups of sugar. You can administer medication in the sugar syrup at this time of year, too. (See Chapter 4 on diseases that affect honey bees.)

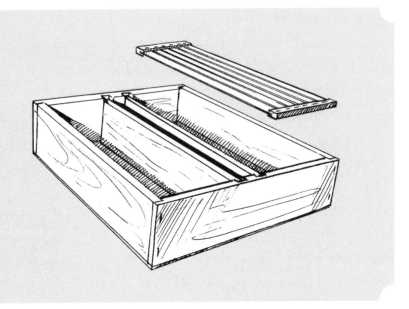

Another way to provide nourishing sugar syrup for your bees is with hive top feeders, which fit under the inner cover.

HUNKERING DOWN *for the* WINTER

November/December

You may want to bundle up your beehives for the winter, and this is a sound instinct—in part. While you certainly want to protect the hive from freezing cold and harsh winds, you also need to provide adequate ventilation. Without it, as the warm air from the cluster rises and reaches the top of the hive with no place to go, it will condense and form drops of cold water that will drop back down into the hive and chill the bees, with often fatal results.

The inner cover features an oval hole for ventilation and through which to feed occasionally. If you place the inner cover so the flat side is down and you expose the notch in its ledge, this can serve as a means of sufficient ventilation—and as an emergency exit, if need be. When you put the outer cover on, though, be sure to push it forward a bit so the hole in the inner cover stays open.

Hopefully when you set them up in the spring, you located your hives in a place that protected them from the wind. If the natural protection of that spot is compromised during the winter, you should reinforce it by building a windbreak. Place some strong posts in the ground and put burlap or other windbreaking material between them. This will greatly reduce the force of the wind blowing against the hive.

You'll want to keep mice from trying to get into the hive, too. The best way to do this (and provide additional insulation) is to reduce the entrance as much as possible and then, for the space that exists, secure some metal screening that has a three-eighths- to half-inch-sized mesh over the open spot. This is large enough for bees to get through, but not mice. Don't think the wood alone will keep mice out. They will just nibble away at the wood until there's a space large enough for them to wiggle through—and that's not a big space.

Beekeepers who live in the northern United States or Canada and other cold places sometimes wrap their hives in thick tar paper as added protection. Some beekeepers feel that this traps moisture, which can be detrimental. This is another time when knowing other local beekeepers or belonging to a beekeeping club can come in handy. Ask several of them what they do. While everyone will have an opinion, you should be able to reach a consensus and at least decide what will be best for you and your hives.

If you choose to wrap the hive, be careful not to block your source of ventilation. You'll also want to put a heavy rock or two on the top of the hive so the paper doesn't blow off.

Winter = Waiting and Wondering

Once you've fed and prepped your hives for the winter, you should leave them alone until spring. You may notice some activity around the hive. The bees don't actually hibernate, so they are still active within the hive and will emerge every once in a while. Some will die, and you may find them near the entrance. Just brush them away and don't worry—they are the older bees who need to die so the younger ones can live on and contribute to repopulating the colony in spring. For the most part, the bees are focused on survival—keeping the queen fed so she can begin laying eggs again in early spring, and feeding themselves by moving around the hive to access stored honey.

And Then, Spring Again!

It may seem like forever between the time you batten down your hive's hatches in late fall and the first time the sun shines on a still day in early spring and sends temperatures into the fifties. When that day comes, however, it is the first sign that the natural world is reawakening in your part of the universe. Hooray!

You can take a quick peek at the hive on a day like the one described above—don't get carried away and risk exposing the bees to temperatures that are too chilly, or to a cold breeze despite a warm sun. Wait for a day when the air is still, the sun is out, and it really is at least 50°F outside. In your protective gear—and with your trusty hive tool and smoker in hand—you can go check the status of your hive.

Gently pry off the cover and take a look inside without moving anything. Do you see the colony? Are there still a lot of bees there? They should be moving around as they are assisting the queen in her early egg-laying days. If all looks well, you can gently remove a frame from near the center of the box. See if you can find the queen, or at least take a look to see if there are eggs being laid. If things look good, put the frame back in the hive box and call it a day.

Start feeding your bees a few weeks before you know things will be blossoming. This will ensure that the bees are getting the energy they need to be as productive as possible. It also allows for a means of medicating the bees, as the medicine can be incorporated into the sugar syrup (more on that in Chapter 4). In addition to the sugar syrup, you can supplement with a pollen substitute until you see that the bees are bringing in pollen from their own trips. The pollen substitute further strengthens the hive and stimulates egg laying by the queen.

If your initial inspection yields an empty box or a lot of dead bees, don't worry too much. Chalk it up to experience, call a fellow beekeeper to commiserate, and start making plans to bring in another package and restart your hobby. This happens to the best of the beekeepers, and you will just need to move on. Prepare for your new package as you did early last year, by getting the hive ready. This year, that will mean cleaning the old hive thoroughly so there are no traces of the last colony.

BEEKEEPER'S YEAR *in* REVIEW

Dick Johnson is a longtime member of the Catskill Mountain Beekeeping Club in New York State (Catskillbees.org). He also writes a column for the *Windham Journal* called "Honeybee Corner." In the January 22, 2009, column, he summarized a beekeeper's activities for the year:

JANUARY: Join a local honey bee club.

FEBRUARY: Attend a beginner's introductory course, read and learn as much as you can about beekeeping, and order your bee supplies and bees as early as possible.

MARCH: When your supplies start coming in, start assembling your hives.

APRIL: Choose the proper location and plan for a bear fence if needed. Try to install new bees around the time that dandelions are blooming.

MAY: Feed new packages, check for good laying queen and parasites. For established hives, watch out for swarms. Watch your bees bring in multicolored pollen from hundreds of fruit trees and spring flowers.

JUNE: Split hives with extra brood, order extra queens. Add new super boxes for honey crop. Sit alongside your hive, watch the activity at the entry, and inhale the delightful aroma of the ripening honey.

JULY: Harvest early honey, extract and return supers for later harvest. Make up splits and nucs for next season. Observe the increased yield of vegetables in your garden as your bees pollinate them.

AUGUST: Take off late honey harvest. Plan for fall disease or parasite management as needed. Do not collect any honey in supers while applying any medication.

SEPTEMBER: Heavy nectar flow from goldenrod and asters may be harvested if no medication is used or else leave on hive for winter colony supplies. Be sure to leave enough honey on each hive for them to eat and keep warm all winter.

OCTOBER: Install entry mouse screens at first frost. Begin plans for winter hive protection. Wind protection and ventilation are needed.

NOVEMBER: Put the hives away for the winter and wait for next spring. Enjoy your first honey harvest and give away or sell your surplus. Continue to attend bee club meetings and share your stores and new fun with bees.

There's no listing for December—gotta love a hobby that gives you a month off!

Chapter 3

Understanding The Life Cycle & Behavior of Honey Bees

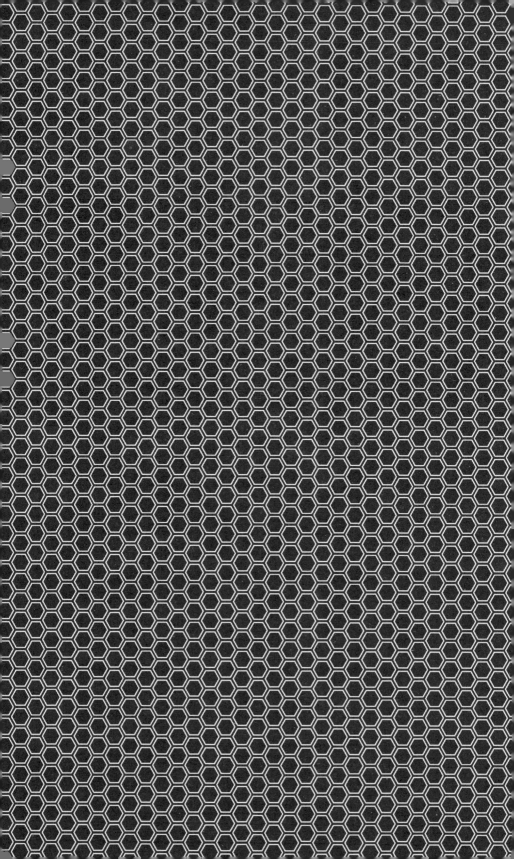

e've already learned that there are three types of bees: workers, queens, and drones. The queen is responsible for laying the eggs that will develop into workers and drones—and into new queens—though it is the workers who identify which eggs they want to feed a special diet so that they become the next generation of queens. All bees follow the life-cycle stages of egg, pupa, larva, and adult, just like butterflies. Each, however, takes a different amount of time to go through those phases.

LIFE CYCLE

Q ueen bees spend all their time going from cell to cell and laying eggs in them. A productive queen can lay two thousand or more eggs a day and sustain a colony of tens of thousands of bees. The cells she chooses have been prepared for the eggs by industrious worker bees who clean them so that they are sterile and spotless. If the cell isn't satisfactory, the queen will move on to another one.

Worker bees also regulate the proportion of workers to drones when they build and clean the cells. Larger cells are for drones and there are far fewer of them. When the queen deposits an egg in a worker cell, it becomes an infertile female worker. The wider cells are reserved for drones, which start out as unfertilized eggs but develop into fertile males. Beekeepers can instantly tell whether their hives are healthy by identifying eggs, pupae, and larvae in cells. The eggs can be hard to spot at first, as they are teeny-tiny, and the queen lays them in an upright position so that each sits at the bottom of a cell.

After three days, regardless of which type of bee they will become, all eggs evolve into larvae. Still tiny, these look like puffy grains of rice. As they grow (quite rapidly) they shed their skins five times. The worker bees feed the larvae as they grow, and after just five days for workers and queens, and seven days for drones, their cells are capped with beeswax. The developing larvae then spin a cocoon around their bodies, and enter the pupa stage of life. The pupae develop all of the features of a full-grown bee as they grow. The pupae have different growth rates depending on which type of bee they will emerge as. The drones take the longest to develop fully—about twenty-four days from the day they are laid as eggs. Worker bees emerge from their cells after twenty-one days. When queen cells are allowed to fully develop, they emerge from their cells in just sixteen days from the time they're laid as eggs. Queens are fed a special diet by the workers so that they emerge fertile and ready to mate with a drone. If the queen is laying eggs, and those eggs are hatching, then the hive is alive and well.

The bees chew their way out of their cells and then go through some additional phases of development. The plentiful worker bees spend their first few days cleaning the area where the eggs are laid. They then go on to feed older larvae with a mixture of honey and pollen. When the worker bees are about seven days old, their pharyngeal glands develop so they can produce royal jelly. They continue to nurture developing eggs with honey, pollen, and royal jelly if the egg is slated to be a queen.

When they are about a week old, worker bees begin taking short flights from the hive. When they are nearly two weeks old, the wax glands on their undersides develop and enable them to start building cell walls and combs. Worker bees do everything: They nurse the young, they fan the hive for ventilation, they process the pollen and nectar that are brought into the hive, they guard the hive against intruders, and they serve the queen bee. At approximately three weeks of age, they begin actively foraging. In another three weeks, when they are about six weeks old, their lives are spent and they die. When they expire, other bees remove them from the hive, taking the corpses some distance away.

The bees that emerge as fertile drones have just one job: to fertilize a young queen. The drones don't have stingers, so they can't defend the hive. They are fed by worker bees until they go out on flights looking for queens. They live about eight weeks, but drones that emerge in late summer may not even live that long, as they are excluded from the cluster of the hive during the winter. When the temperatures drop and the colony moves in on itself to cluster for the winter, young worker bees will actively chase drones out of the hive, sending them to their frosty deaths.

Queen bees who survive long enough to emerge from their cells have an instant following of worker bees, who tend to their every function as they establish a new colony. The success of an entire hive depends on the genetic makeup of its queen bee, as her genes will be in every egg that is laid in the hive. During a particular year, the bees that are born to the hive may have the genetic makeup of any number of male bees along with that of the queen bee, as her mating flight enables her to collect six to twelve million sperm to use throughout her life. To mate, a virgin queen leaves the hive on six or seven consecutive days and flies to a particular area that is shared by several colonies. The drones go there, too, and mating occurs in midflight.

ON *the* BEHAVIOR *of* BEES

There are excellent books, essays, poems, and other writings about bees. (You can find their titles in the Beekeeping Resources section of this book.) There are even wonderful movies about them. In the earlier chapters, the writer John Burroughs is quoted in several places. He lived near the Catskill Mountains in New York, and wrote extensively about the natural world as he observed it. Fortunately for us, one of his favorite subjects was the honey bee. If you're looking for an understanding of the nature of *Apis mellifera*, this passage from "The Pastoral Bee," from *Birds and Bees*, written in the late 1800s, is truly inspirational:

> The honey bee goes forth from the hive in spring like the dove from Noah's ark, and it is not till after many days that she brings back the olive leaf, which in this case is a pellet of golden pollen upon each hip, usually obtained from the alder or the swamp willow. In a country where maple sugar is made, the bees get their first taste of sweet from the sap as it flows from the spiles, or as it dries and is condensed upon the sides of the buckets. They will sometimes, in their eagerness, come about the boiling place and be overwhelmed by the steam and the smoke. But bees appear to be more eager for bread in the spring than for honey; their supply of this article, perhaps, does not keep as well as their stores of the latter, hence fresh

bread, in the shape of new pollen, is diligently sought for. My bees get their first supplies from the catkins of the willows. How quickly they find them out. If but one catkin opens anywhere within range, a bee is on hand that very hour to rifle it, and it is a most pleasing experience to stand near the hive some mild April day and see them come pouring in with their little baskets packed with this first fruitage of the spring. They will have new bread now; they have been to mill in good earnest; see their dusty coats, and the golden grist they bring home with them.

When a bee brings pollen into the hive, he advances to the cell in which it is to be deposited and kicks it off as one might his overalls or rubber boots, making one foot help the other; then he walks off without ever looking behind him; another bee, one of the indoor hands, comes along and rams it down with his head and packs it into the cell as the dairymaid packs butter into a firkin.

The first spring wild-flowers, whose shy faces among the dry leaves and rocks are so welcome, yield no honey. The anemone, the hepatica, the bloodroot, the arbutus, the numerous violets, the spring beauty, the corydalis, etc., woo lovers of nature, but do not woo the honey-loving bee. It requires more sun and warmth to develop the saccharine element, and the beauty of these pale striplings of the woods and groves is their sole and sufficient excuse for being. The arbutus, lying low and keeping green all winter, attains to perfume, but not to honey.

The first honey is perhaps obtained from the flowers of the red maple and the golden willow. The latter sends forth a wild, delicious perfume. The sugar maple blooms a little later, and from its silken tassels a rich nectar is gathered. My bees will not label these different varieties for me as I really wish they would. Honey from the maples, a tree so clean and wholesome, and full of such virtues every way, would be something to put one's tongue to. Or that from the blossoms of the apple, the peach, the cherry, the quince, the currant—one would like a card of each of these varieties to note their peculiar qualities. The apple-blossom is very important to the bees. A single swarm has been known to gain twenty pounds in weight during its continuance. Bees love the ripened fruit, too, and in August and September will suck themselves tipsy upon varieties such as the sops-of-wine.

The interval between the blooming of the fruit-trees and that of the clover and the raspberry is bridged over in many localities by the honey locust. What a delightful summer murmur these trees send forth at this season. I know nothing about the quality of the honey, but it ought to keep well. But when the red raspberry blooms, the fountains of plenty are unsealed indeed; what a commotion about the hives then, especially in localities where it is extensively cultivated, as in places along the Hudson. The delicate white clover, which begins to bloom about the same time, is neglected; even honey itself is passed by for this modest colorless, all but odorless flower. A field of these berries in June sends forth a continuous murmur like that of an enormous hive. The honey is not so white as that obtained from clover but it is easier gathered; it is in shallow cups while that of the clover is in deep tubes. The bees are up and at it before sunrise, and it takes a brisk shower to drive them in. But the clover blooms later and blooms everywhere, and is the staple source of supply of the finest quality of honey. The red clover yields up its stores only to the longer proboscis of the bumble-bee, else the bee pasturage of our agricultural districts would be unequaled. I do not know from what the famous honey of Chamouni in the Alps is made, but it can hardly surpass our best products. The snow-white honey of Anatolia in Asiatic Turkey, which is regularly sent to Constantinople for the use of the grand seignior and the ladies of his seraglio, is obtained from the cotton plant, which makes me think that the white clover does not flourish these. The white clover is indigenous with us; its seeds seem latent in the ground, and the application of certain stimulants to the soil, such as wood ashes, causes them to germinate and spring up.

The rose, with all its beauty and perfume, yields no honey to the bee, unless the wild species be sought by the bumble-bee. Among the humbler plants, let me not forget the dandelion that so early dots the sunny slopes, and upon which the bee languidly grazes, wallowing to his knees in the golden but not over-succulent pasturage. From the blooming rye and wheat the bee gathers pollen, also from the obscure blossoms of Indian corn. Among weeds, catnip is the great favorite. It lasts nearly the whole season and yields richly. It could no doubt be profitably cultivated in some localities, and catnip honey would be a novelty in the market. It would probably partake of the aromatic properties of the plant from which it was derived.

Buckwheat honey is the black sheep in this white flock, but there is spirit and character in it. It lays hold of the taste in no equivocal manner, especially when at a winter breakfast it meets its fellow, the russet buckwheat cake. Bread with honey to cover it from the same stalk is double good fortune. It is not black, either, but nut-brown.

How the bees love it, and they bring the delicious odor of the blooming plant to the hive with them, so that in the moist warm twilight the apiary is redolent with the perfume of buckwheat.

Yet evidently it is not the perfume of any flower that attracts the bees; they pay no attention to the sweet-scented lilac, or to heliotrope, but work upon sumach, silkweed, and the hateful snapdragon. In September they are hard pressed, and do well if they pick up enough sweet to pay the running expenses of their establishment. The purple asters and the golden-rod are about all that remain to them.

It is the making of the wax that costs with the bee. As with the poet, the form, the receptacle, gives him more trouble than the sweet that fills it, though, to be sure, there is always more or less empty comb in both cases. The honey he can have for the gathering, but the wax he must make himself—must evolve from his own inner consciousness. When wax is to be made the wax-makers fill themselves with honey and retire into their chamber for private meditation; it is like some solemn religious rite; they take hold of hands, or hook themselves together in long lines that hang in festoons from the top of the hive, and wait for the miracle to transpire. After about twenty-four hours their patience is rewarded, the honey is turned into wax, minute scales of which are secreted from between the rings of the abdomen of each bee; this is taken off and from it the comb is built up. It is calculated that about twenty-five pounds of honey are used in elaborating one pound of comb, to say nothing of the time that is lost.

But honey without the comb is the perfume without the rose—it is sweet merely, and soon degenerates into candy. Half the delectableness is in breaking down these frail and exquisite walls yourself, and tasting the nectar before it has lost its freshness by the contact with the air. Then the comb is a sort of shield or foil that prevents the tongue from being overwhelmed by the shock of the sweet.

It is a singular fact that the queen is made, not born. If the entire population of Spain or Great Britain were the offspring of one mother, it might be found necessary to hit upon some device by which a royal baby could be manufactured out of an ordinary one, or else give up the fashion of royalty. All the bees in the hive have a common parentage, and the queen and the worker are the same in the egg and in the chick; the patent of royalty is in the cell and in the food; the cell being much larger, and the food a peculiar stimulating kind of jelly.

In certain contingencies, such as the loss of the queen with no eggs in the royal cells, the workers take the larva of an ordinary bee, enlarge the cell by taking in the two adjoining ones, and nurse it and stuff it and coddle it, till at the end of sixteen days it comes out a queen. But ordinarily, in the natural course of events, the young queen is kept a prisoner in her cell till the old queen has left with the swarm. Later on, the unhatched queen is guarded against the reigning queen, who only wants an opportunity to murder every royal scion in the hive. At this time both the queens, the one a prisoner and the other at large, pipe defiance at each other, a shrill, fine, trumpet-like note that any ear will at once recognize. This challenge, not being allowed to be accepted by either party, is followed, in a day or two by the abdication of the reigning queen; she leads out the swarm, and her successor is liberated by her keepers, who, in her time, abdicates in favor of the next younger. When the bees have decided that no more swarms can issue, the reigning queen is allowed to use her stiletto upon her unhatched sisters. Cases have been known where two queens issued at the same time, when a mortal combat ensued, encouraged by the workers, who formed a ring about them, but showed no preference, and recognized the victor as the lawful sovereign.

The peculiar office and sacredness of the queen consists in the fact that she is the mother of the swarm, and the bees love and cherish her as a mother and not as a sovereign. She is the sole female bee in the hive, and the swarm clings to her because she is their life. Deprived of their queen, and of all brood from which to rear one, the swarm loses all heart and soon dies, though there be an abundance of honey in the hive.

The common bees will never use their sting upon the queen; if she is to be disposed of they starve her to death; and the queen herself will sting nothing but royalty— nothing but a rival queen.

The bees do not fall down before her, but caress her and touch her person. There is but one fact or incident in the life of the queen that looks imperial and authoritative. Huber relates that when the old queen is restrained in her movements by the workers, and prevented from destroying the young queens in their cells, she assumes a peculiar attitude and utters a note that strikes every bee motionless, and makes every head bow; while this sound lasts not a bee stirs, but all look abashed and humbled, yet whether the emotion is one of fear, or reverence, or of sympathy with the distress of the queen mother, is hard to determine. The moment it ceases and she advances again toward the royal cells, the bees bite and pull and insult her as before.

A MORE SCIENTIFIC APPROACH

Scientists are always striving to learn more about honey bee behavior—particularly in light of the recent global epidemic of colony collapse disorder (CCD). Here are some basic ways to understand why honey bees do what they do.

As discussed earlier, the worker bees import into the hive nectar and pollen essential to making honey. They will have reached a certain stage of development before they can make extended trips from the hive and back. When they start leaving the hive, their first trips are short, because they are learning from the older bees.

Older worker bees are foraging experts, having made many trips to and from the hive. They don't want to waste their time—or that of their fellow hivemates—returning to places where there is little in store for them. Instead, they find the plants that are worth the trip, sometimes going back and forth several times during a single day.

The SMELL or the DANCE?

When worker bees find something really good, they communicate it to the other worker bees, who will join the bee that has just relayed this information. Aristotle makes a note of this in his book *Historia Animalium* dating to 330 BC, and suggests that a dance is done to attract the attention of other foragers, who then follow the bee to the source.

While studies of the communication habits of bees were ongoing through the ages, it was Karl von Frisch's work that sparked modern beekeeping. Frisch, an Austrian ethologist who taught zoology at the University of Munich in Germany, noted in the 1920s that scent appeared to be a determining factor for the recruitment of additional foraging worker bees to particular plants. If a linden tree was a great source of nectar, a foraging bee would return to the hive and do a dance that would appear to rally other workers so that they could find the source, too. Frisch observed that a particularly good or fresh source seemed to elicit a more frenzied dance, while dances that weren't so lively indicated a less fresh source or a source that was a greater distance from the hive. He also found that bees that were rallied to forage for a particular nectar (clover, for example) would not respond to the dances of bees who were recruiting for another source (linden, for example).

Frisch continued to study the behavior of honey bees, and in the late 1940s realized that there was more going on in the communication of the bees than mere scent association. So instrumental was his work that he was awarded the Nobel Prize in physiology and medicine in 1973. In his Nobel lecture, "Decoding the Language of Bees," Frisch noted, in part, that communications are "marked by tail-wagging dance movements and simultaneously toned buzzing," with longer distances expressed by longer tail-wagging times. The tail-wagging, according to Frisch, indicates distance and gives the direction to the nectar source in relation to the position of the sun. The dance is performed "in the darkness of the hive, on the vertical surface of the comb, as an angular deflection from the vertical. The bee thus transposes the angle to a different area of sense perception" because bees perceive polarized light and can identify the sun's position even when it is obscured by a mountain. (Frisch's entire Nobel lecture can be found at Nobelprize.org by searching for "Karl von Frisch.")

While many researchers believe that bee dances give enough information to locate resources, proponents of the odor-plume theory argue that the dance gives no actual guidance to a source of nectar. They argue that the purpose of the dance is simply to gain attention for the returning worker bee so she can share the odor of the nectar with other workers. For them it is the odor that dictates the source. This theory resulted from experiments that used odorless sugar sources. In those trials, worker bees were unable to recruit others to the nectar source, despite the source being nearby and easily accessible. You can watch bees do the "waggle dance" online by going to youtube.com and searching for "honeybee dance." Wow!

PHEROMONES *and* BEE-HAVIOR

Pheromones (derived from the Greek words meaning "to bear" and "hormone") are chemical combinations producing scents that animals give off to other animals of the same species. The pheromones provide information and produce behavioral responses. They are transmitted by glands and received by the bees' antennae and other body parts. Honey bees utilize a number of pheromones, which are instrumental in all aspects of colony life; in fact, honey bees have one of the most complex pheromonal communication systems found in nature. Consider the following types of pheromones utilized by honey bees:

······ QUEEN PHEROMONES ······

The pheromones produced by queen bees are essentially what differentiate queen bees from worker and drone bees. Queen pheromones stimulate all the activities of the worker bees, from brood rearing to comb production to foraging and more. The queen's pheromones also inhibit ovary production in worker bees, rendering the workers infertile, and are responsible for swarming and mating behaviors.

······ ALARM PHEROMONES ······

These are produced by worker bees for all manner of defensive or aggressive behavior. When a bee stings, a pheromone directs other bees to sting in the same place. Alarm pheromones are given off between robbers invading a hive, and among hive dwellers being attacked.

······ BROOD RECOGNITION PHEROMONES ······

Developing pupae and larvae emit pheromones to help nurse bees differentiate workers from drones. These pheromones also inhibit ovarian development in worker bees.

······ DRONE PHEROMONES ······

Drones attract other drones to a mating location via drone pheromones. After the drones are in place, a virgin queen can be attracted and, in one remarkable session, fertilized for her lifetime.

······ NASONOV PHEROMONES ······

The sweet-smelling Nasonov pheromones are released from a gland at the end of each worker bee's abdomen. They help direct foraging workers back to their hive (and not to the hive of other bees). The etymology of the name is uncertain.

······ OTHER LESSER PHEROMONES ······

Bees use many other pheromones to communicate, such as the footprint, forager, rectal gland, tarsal, wax gland and comb, and tergite gland pheromones. The more you observe and get to know them, the more fascinating and amazing honey bees become. If the behavior of bees becomes a passion of yours, there are some wonderful books about it. (See the Resources section.)

Chapter 4

Keeping
Bees Healthy

oney bees, like all creatures, can become diseased, and the best way to combat illness is to take a preventive approach. Knowing what can affect your bees, and then using careful observation to identify a developing problem, is the best way to ward off disease in your hives.

Honey bees are susceptible to certain diseases and to certain parasites. They are also subject to colony collapse disorder, which will be discussed in more detail at the end of the chapter.

According to the Dyce Laboratory for Honeybee Studies at Cornell University, pest management is a sound approach to disease management, because it enables you to minimize or even eliminate the use of antibiotics. This has two highly desirable consequences, according to the lab. First, it reduces the likelihood that the pathogens will develop antibiotic resistance. Second, it reduces or eliminates antibiotic residues in your honey. This, in turn, allows you greater access to the natural and organic foods markets and to the higher prices they offer at both producer and retail levels.

According to Dyce Laboratory, an effective IPM (Integrated Pest Management) program for disease management has three basic requirements:

1. You must be able to accurately identify the major diseases.

2. You must know what to do when you encounter a disease.

3. You must incorporate basic disease management protocols into your overall management scheme.

Inquire about any pest management workshops local beekeeping organizations conduct or review any literature they have available. People in the area are your first line of education and defense; they're most likely to know what your bees are particularly susceptible to. Beekeepers who live nearby can also give you good advice on when to treat in the spring and fall if this is part of their normal disease prevention protocol. This chapter gives an overview of the most common diseases of honey bees.

PREVALENT BEE DISEASES

There are five diseases that most often harm honey bees. They are: American foulbrood, European foulbrood, chalkbrood, sacbrood, and nosema. Let's take a closer look at each one.

American Foulbrood (AFB)

American foulbrood is a highly contagious bacterial disease that affects honey bee larvae. Once infected, the entire colony can be wiped out if the disease goes untreated.

AFB has two stages in its life cycle—one is a vegetative or rod stage, the other is a long-lived spore stage. The spores are typically fed to developing larvae by infected nurse bees. Once in the larvae themselves, the spores germinate to a rod stage, reproducing rapidly, spreading through the larvae and killing them. When the larvae die, the rods become spores, which can remain dormant in the hive for more than thirty years. The spores are spread when worker bees go into the cells of the dead larvae to clean them out.

Symptoms

As the larvae are taken over by AFB, they turn from a pearly white color to dark brown. The tissues harden and scale, and it is the scales (which contain the spores) that attach to the worker bees and in turn get passed through the hive. When inspecting your frames, if you notice any of the following symptoms alone or in combination, you should suspect AFB and act quickly to eradicate it. Look for:

- An irregular brood pattern.

- Larvae changing in color from white to brown.

- A foul odor.

- Capped cells darkening and appearing sunken rather than convex. Cells may also be moist.

- Dried scales with something sticking out of them and into the bottom of the cell.

If the cells look diseased to you, you can take your inspection a step further by using a toothpick to poke into a cell. Give the toothpick a twirl, then slowly pull it out. If the larva is infected, it will have a stringy, taffylike appearance and texture. When you pierce the cap, you will release the odor, too, and it is distinctive.

Treatment

Because of the contagious nature of AFB, treatment needs to be immediate. At the first sign, apply the antibiotic Terramycin (also known by the acronym TM) as a syrup, dust, or food patty. Dosage instructions are indicated on Terramycin packages, which are available through bee supply companies. Many beekeepers routinely feed TM to their hives in the spring and fall to help prevent an infection. Because of its aggressive nature, it doesn't take long for AFB to seriously impact a hive, at which point you will need to consider destroying whatever may be infected. Never apply TM to colonies with active AFB or AFB scales. AFB-infected colonies must be treated according to local regulations. Usually, this requires that the colony be killed and the associated equipment be burned or buried.

European Foulbrood (EFB)

Not as common as AFB, European foulbrood isn't as destructive, either—but beekeepers know that it can still cause a lot of damage, and that simple preventive measures are worth every expense of time and money. EFB is caused by the bacteria *Streptococcus pluton*. There is no spore form. It infects larvae when they are just a few days old, and they die before they are capped in their cells. Often the larvae will rid themselves of the infection by defecating the infection out of their bodies. If and when they do this, the feces can remain in the comb and risk infecting it for several years.

Symptoms

As the larvae are infected by EFB, they turn from a pearly white color to light tan. While normal larvae are appropriately moist and glistening, those infected with EFB appear to have a smooth, sleek, almost slimy appearance, and they twist into a corkscrew shape in the cell. When inspecting your frames, if you notice any of the following symptoms alone or in combination, you should suspect EFB and act quickly to eradicate it. Look for:

• Coiled larvae in the cells.

• Larvae that are pale yellow to tan and sleek rather than just moist in appearance.

• A foul smell, though not as strong as AFB odor.

Treatment

Just as with AFB, EFB is treated with the antibiotic Terramycin, applied as a syrup, dust, or pellet. Dosage directions provided on the label should be strictly adhered to, and many beekeepers give a preventive treatment in the spring (along with the one for AFB). If infection is serious, a sample should be sent to your local agricultural extension office. Test results will determine whether or not the hive may need to be destroyed.

Chalkbrood

Chalkbrood is a fungal disease that affects the larvae of worker and drone bees, the most prevalent honey bees in a hive. Fortunately, it rarely affects an entire colony. The fungus is *Ascosphaera apis*, and there are two forms of it—the vegetative form and a spore form. Just as with AFB, the spores can remain viable for years.

Symptoms

Chalkbrood typically manifests in the spring and is believed to be precipitated by cool, damp weather. A fungal growth in a honey bee brood will be similar to one on old food or a rotten vegetable—it becomes covered with a white, fluffy substance. In larval cells, this white material fills the entire cell and then dries, forming larval mummies that are easily removed from the cell.

If the fungus assumes the spore form, it can cause the mummy to become mottled in black and white bumps or become completely black. A heavily infected colony will have a lot of mummies at its entrance or in capped cells throughout the brood.

Treatment

There are currently no approved medications for control of chalk-brood, so it's fortunate that the disease is usually self-limiting. If your hive is infected, remove the mummified carcasses both in and around the hive, assisting the worker bees with cleanup. If one of your frames seems particularly infected, remove it and replace it with a new, clean frame. Making sure your hives have plenty of sun in the springtime is helpful, too.

Sacbrood

Should this viral disease infect your bees, you'll notice the larvae turning color—from pearly white to yellow to dark brown—and the cell filling with fluid. Worker bees usually spot the infected cells and remove the larvae, limiting the spread of infection. There are no approved medications for sacbrood.

Nosema

The other diseases discussed here affect the larvae, but nosema affects adult honey bees. It typically manifests in the spring when the bees become active again. Often, when nosema strikes, it infects the queen, leading to the need to requeen the hive.

Nosema is caused by a single-celled protozoan with two growth stages. The first is an actively growing cell stage that reproduces in the bees' midgut. The second is a spore stage that can remain viable for years and continue to infect a hive as it gets passed by other bees when they defecate.

Symptoms

A nosema infection results in bees who are extremely weakened, as they are unable to process food and don't live long enough to do all their work. A bad infection can impact honey production by 40 to 50 percent. Infected bees appear to stagger around, and their overall production is limited. Their abdomens can appear distended, as the infection is in the midgut. You will sometimes see streaks of light-colored feces in and on the hive.

Treatment

Beekeepers aim to prevent a nosema infection by treating their bees with the antibiotic fumagillin or fumidil. As with all medications, the dosage directions must be strictly adhered to. Other preventive measures include locating the hives where they will be adequately ventilated and where there is access to clean, fresh water.

PARASITES *that* AFFECT HONEY BEES

Honey bees are hosts to a number of parasites that can seriously infect and weaken a hive. Keeping an eye on these and following preventive measures is as critical as doing what you can to prevent disease. At the top of the "pest" list are two mites—the varroa mite and the tracheal mite. Howland Blackiston, author of *Beekeeping for Dummies*, sums it up this way: "Doing nothing to protect your bees from mites is like playing a game of Russian roulette." Ouch!

Varroa Mites

These pests originated in Asia and are now found on honey bees around the world. They were first spotted by American beekeepers some thirty years ago—and they've been fairly relentless ever since.

Varroa mites (*Varroa jacobsoni*) are visible to the naked eye when they are on the honey bees. They are about the size of a pinhead, and typically bees have several on them at a time. The ones that are visible are the females, whereas the male adults and nymph stages of the mites are smaller and are yellowish to grayish white. Adult females feed not only on the bee but on the brood, and the males remain sealed inside the brood cells with the developing larvae.

The life cycle of the varroa mite begins as an egg laid by a female mite that has crawled into the brood cell of a drone or worker bee (preferably a drone). When the cell is capped, the female mite lays several eggs, which feed on the developing bee. Female mites take eight to ten days to develop; males take just six to seven days. The males mate with their sisters and then die in the cell. The mated female mites emerge with the hatching bees, already attached to them. They instantly seek a blood meal on an adult honey bee, and are then ready to go back into a cell to lay their eggs, perpetuating the cycle.

The damage that varroa mites do to honey bees is to weaken them. By feeding on their blood, they can cause shortened life spans, shortened abdomens, deformed wings, and bees who weigh less than they should. Though the mites prefer drone cells and hatch from them, they are quick to feed on worker bees, who seem to suffer more than the drones. Wounds caused by the bites of varroa mites are susceptible to infection, further weakening the bees.

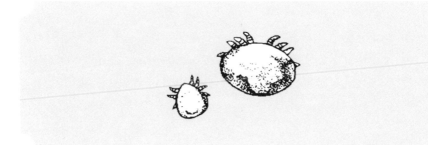

Illustrated above are magnified varroa mites. In real life, they are the size of a pinhead. Since the mite is such a prolific pest of honey bees around the world, there are proven ways to prevent or cure infestation.

Symptoms

Hives with low infestation rates may appear to be healthy as there is a far lower rate of incidence. Beekeepers need to beware and not let down their guard against varroa mites, as by the time there is a heavy infestation, the situation is not only serious, but also harder to handle. Because varroa mites prefer to lay their eggs in drone cells, beekeepers must assess the actual cells themselves. This should start in the spring and continue through the season, using the following steps as a guide:

1. Use a cap scratcher to get past the caps of the cells and lift out the pupae and the caps.

2. Use forceps to scratch off the cap so the pupa can be lifted out of the cell for observation.

3. Use a small, sharp knife to cut off several drone cell caps, then tap the frame against a hard, clean, white surface to observe any pupae that fall out.

4. Take a cross-examination of cells throughout the hive and observe pupae and cell bottoms, looking specifically for the mite eggs.

Because the developing pupae are white, once they are removed for observation it is easy to see the varroa mites on them.

A varroa infestation can be detected in other ways, too.

- Emerging bees appear deformed—especially their wings or abdomens.

- You see small brown spots on white larvae.

- You see pinhead-sized brown spots on adult bees (at which point your hive is probably heavily infested).

Treatment

There are several methods available for treating your honey bees for varroa mites. Again, it is so helpful to be involved with a local beekeeping club, as its members will certainly have had experience with the mites and with different treatment methods. There may be some preferences among members and you can avail yourself of their experience.

APISTAN STRIPS: The miticide used in the Apistan anti-varroa strips is fluvalinate. The strips look like bookmarks, and per the directions—which should be followed carefully—they should be hung in the brood chambers between the second and third frames and also between the seventh and eighth frames. The positioning ensures that the bees will come into contact with the strips as they move about the hive. Once in place, the strips need to be left in the hive for the amount of time specified in the directions—neither shorter nor longer, as that could render the treatment ineffective.

Apistan strips can only be used when honey supers are off. They should be handled with gloves so that the miticide does not come into contact with your skin. Temperatures should be steadily in the fifties (Fahrenheit) during the day when Apistan is used, and the strips cannot be left in the hive over the winter.

Some recommend enhancing the effectiveness of Apistan strips in the hive with detector boards (screened bottom boards). These can be constructed from white cardboard fitted to the hive's bottom board, over which is added a screened cover. The size of the screen should be large enough for mites to pass through but small enough for the bees to not be able to go through—a size 8 mesh is appropriate. With the cardboard on the bottom board, apply a light layer of vegetable oil or petroleum jelly—something to which the mites will stick when they drop through the screen so that they can't climb back up into the hive. Place the screened board securely over the bottom board with about a quarter inch of space.

With normal hive activity, the cardboard on the bottom board is quickly coated with debris, some of which includes the mites. You should check and replace this frequently. The detector boards can be used with or without Apistan strips (or other miticides), but are especially helpful when treating a hive, as you'll be able to see whether mites are indeed falling off the bees.

Treatments and Honey Production

It is illegal for pesticides to be present in a colony during a marketable honey flow. Treatment with Apistan requires a minimum of forty-two days; therefore, you must install strips at least forty-two days before adding supers for honey production. Treatment with CheckMite+ also requires a minimum of forty-two days, but you must begin treatment at least fifty-six days before adding supers for honey production.

CHECKMITE+: This is a relatively new miticide developed as a result of bees' increasing resistance to fluvalinate. The chemical used in this product is the miticide coumaphos—which is also used in nerve gas. CheckMite+ comes in strips that are placed in the hive in the same way as Apistan strips. CheckMite+ is heavy-duty stuff that should really only be considered if a hive is known to be resistant to Apistan.

Tracheal Mites

Where the varroa mite is an external parasite, the tracheal mite is an internal parasite—it does its damage from the inside out. These small but deadly pests (*Acarapis woodi*) spend their entire life cycle inside the bee—and even if they were external parasites, they are so small that they could not be detected by the naked eye.

The life cycle of the tracheal mite begins when an adult female lays several eggs in the trachea (breathing tubes) of an adult bee. The eggs hatch over the course of four days and in four stages: egg, larva, nymph, and adult. The larval, nymphal, and adult stages of the mites all have mouthparts that pierce the tracheal tissues so they can feed on the bee's blood. The mites mate in the trachea and then die. Peak populations occur in the winter.

Symptoms

Because these pests can only be detected under a microscope, any suspicion you have about an infestation needs to be confirmed by a state apiary inspector. The local beekeeping club will know how to contact this person if you aren't sure. The clues that you might have a tracheal mite infection in the hive aren't always conclusive, so it's best to err on the side of caution and have the apiary inspector come by to give you a definitive answer. Clues may include:

• Bees appear sick, weak, or disoriented near the entrance to the hive or on the ground. Because the mites clog the breathing tubes, infected bees are extremely weakened. If you see bees trying to climb up pieces of grass or plant matter to try to fly but then not being able to take off, they are being weakened by something.

- Bees have "K-wings" that appear "unhooked" or are extended at odd angles.

- Small or split clusters are sometimes indicative of tracheal mites. If there is plenty of honey and bees are either dying in the winter or abandoning the hive, suspect tracheal mites.

Treatment

If there's a piece of good news when it comes to tracheal mites, it's that the treatment for them is a natural product rather than a potent chemical. Beekeepers use menthol to fight against tracheal mites. The menthol comes in prepackaged bags of crystals.

Before using, remember to remove any honey supers; and, though the product is natural, it still needs to be used according to the package directions. Packets are placed over the bars at the top of the brood chamber, toward the rear of the hive. A piece of aluminum foil should be placed under the packet so any dissolved liquid doesn't trickle down into the hive. It's the menthol vapors that need to be dispersed and which the bees breath in so the mites are killed. So that it warms and properly dissipates, outdoor temperatures should be between 60 and 80 degrees Fahrenheit, and the packets need to be left for a specific amount of time.

Beekeepers Bob Noel and Jim Amerine began experimenting with essential oils to help control mites and developed a product called Honey-B-Healthy. Combining spearmint and lemongrass oils along with other natural ingredients, Honey-B-Healthy is a food supplement that can be added to the sugar syrup that's given to bees on a regular basis.

Other Pests *and* Problems

Wax Moth

There are two types of wax moths: the greater wax moth, and the lesser wax moth. Not surprisingly, the greater wax moth does the greater damage to hives, potentially ravaging combs. While it is usually the adult that is visible to a beekeeper, it is the larval or caterpillar stage that does the damage—especially to combs that are removed from a hive and kept in storage. If combs are stored in dark, warm, or closed-in areas, wax moths move in quickly and can decimate the combs.

Life Cycle

There are four stages in the life cycle of wax moths: egg, larva, cocoon, and adult. Eggs are laid by adults in between hive parts. They are tiny and difficult to detect. Females can lay three hundred eggs each. When they hatch, the larvae burrow into the beeswax comb, where they feed on impurities in the comb. They prefer comb that has been used for brood rearing. The larval stage can last from one to five months, depending on conditions. As they grow, the larvae are visible to the naked eye and look like pale gray caterpillars.

At the cocoon stage, the larvae encase themselves in dense and tough cocoons that attach themselves to the bottom of the hive or, frequently, on the frame or body of the hive so that they bore into the wood. Infestations often mean having to trash the whole apparatus. The pupal cocoon stage is relatively short, and adults emerge as moths. They grow to be up to one and a half inches long and are a pale gray color, with males slightly smaller than females.

Treatment

The bees themselves make every effort to rid their hives of wax moths, and will move the larvae or caterpillars to places where there is little for them to feed on. Weak colonies can't do as good a job, and because the female moths lay so many eggs, an infestation can set in. If you don't notice the eggs, you may notice that healthy-looking combs are darker and no longer house brood.

Beekeepers can assist their bees in keeping hives free of wax moths by helping to keep the hives clean and adequately ventilated so that bees have access to all parts of the hive. Paying regular attention to brood-bearing combs is critical, too.

The time that beekeepers need to be most vigilant about wax moths is when they're storing the honey supers after collecting the honey.

Preventive measures like putting the combs in plastic bags or leaving them in a colder environment don't help much, as an existing problem can fester in either condition. The best thing to do is treat the supers with a moth fumigant. Many bee supply companies sell PDB (paradichlorobenzene) crystals, and this is the fumigant of choice.

When you're ready to store your supers for the winter, follow the directions for the amount of PDB crystals you should use, and be sure that you have some between each super so the vapors can spread throughout. Put them on a small piece of cardboard so they're not directly on the bars of the super. Stop up any air holes and put a top on the stack of supers so that they're protected. When you take them back out in the spring, it is critical to separate and air them for at least forty-eight hours before putting them back in the hive.

One Additional Pest

Where temperatures are consistently warmer, hives can come under attack by the small hive beetle. While beetles in general are fairly harmless to bees, this one thrives on eating everything associated with honey bees—honey, pollen, wax, combs, brood, eggs. The little black beetles are visible to the naked eye, though they are small. Any suspicion of a small hive beetle invasion should be confirmed by a state apiary consultant. Treatment is typically with CheckMite+.

Bears, Varmints, and Other Troublemakers

In their search for honey, bears will invade an apiary and basically ransack it. You can observe a lovely setup one summer evening and come out the next morning to find your hive boxes tossed about like they'd been in a tornado—all while you slept soundly. If the sight of your hives in pieces isn't upsetting enough, imagine what your bees had to endure.

The bear is a major predator to a beekeeper's hives, with the potential to cause a lot of destruction. A bear fence is necessary in areas with known bear populations.

To protect against this predator, beekeepers in any areas where bears are known to live must take preventive measures early on and, while they're establishing their hives, encircle them with an electric fence to (hopefully) keep out the predators. This is something you'll want to talk to your local beekeeping club about when you're getting started. You may not think it's necessary, but when all of a sudden your hives are victims of a raid, you'll wish you'd put up a fence. It depletes your apiary of a certain rustic charm, but it can be well worth it.

Depending on where you live, your bees may be a target for skunks, raccoons, possums, mice, and other small animals. While being stung may send some of these smaller, thinner-skinned animals running, others are persistent. You can further deter their interest in the hive by making it less accessible for them. This means minimizing the bees' entrance so that the other animals can't make their way inside, or further protecting the hive by elevating it and putting some chicken wire in front of the entrance.

Mice are especially troublesome, as they can sneak into very small spaces. They're not so much a threat to the bees themselves, but they can seriously damage the hive. They tend to come into hives in the fall and find a corner in which to build a nest. The material for their nest is gnawed wood from the hive or frames. They also urinate and defecate in the hive, making it stinky and the bees reluctant to use the soiled area(s). Beekeepers need to be especially diligent about preventing mice from getting into the hives at summer's end by reducing the size of the entrance.

Ants

Ants love sugar, and a hive is essentially a giant sugar cube to them. Bees naturally guard against curious and hungry ants, and manage to keep them at bay. A healthy, thriving colony does an excellent job with this. But a young or weakened colony may have trouble keeping up, and if it feels under attack, the colony may even leave the hive for a new home.

If you notice what seems to be an unusual amount of ants around or in the hive, it's time to take action. One of the most effective preventive measures is to raise the hive off the ground, elevating it onto strong posts at each corner so that the hive is lifted about eighteen inches off the ground. Secure the legs of the posts in tin cans and fill the cans with motor oil. This will not hurt the bees, and the ants can't climb up and cross the "moat" of oil.

COLONY COLLAPSE DISORDER: *A* COMPLEX BUZZ

The USDA Agriculture Research Service offers the following information about a recent but devastating development in beekeeping:

In the fall of 2006, a loud, new buzz began among beekeepers in a number of countries when managed honey bee colonies began to disappear in large numbers without known reason. By February 2007, the syndrome, which is characterized by the disappearance of all adult honey bees in a hive while immature bees and honey remain, had been christened "colony collapse disorder" (CCD).

Some beekeepers reported losses of 30 to 90 percent of their hives during the 2006 winter. While colony losses are not unexpected during winter weather, the magnitude of loss suffered by these beekeepers was highly unusual.

Because honey bees are critical for agricultural pollination—adding more than $15 billion in value to about 130 crops—especially high-value specialty crops like berries, nuts, fruits, and vegetables, the unexplained disappearance of so many managed colonies was not a matter to take lightly.

In general, honey bee colony health has been declining since the 1980s, with the introduction of new pathogens and pests. The spread into the United States of varroa and tracheal mites, in particular, created major new stresses on honey bees. At the same time, the call for hives to supply pollination services has continued to climb. This means honey bee colonies are trucked farther and more often than ever before, which also stresses the bees.

While CCD is truly a serious problem, agricultural pollination is not in crisis at this time. There were enough honey bees to provide all the pollination needed in 2007. Specific reports of CCD during [2005] were not greater than they were in 2006. But a survey of managed hives done in fall and winter 2007 by the Bee Research Lab and the Apiary Inspectors of America showed that beekeepers lost about 35 percent of their hives compared to 31 percent in 2006, so bee losses overall are not improving.

The new syndrome may have a name—CCD—but the question for beekeepers and scientists alike is, "Just what is causing CCD?"

"It's a very good question. And everyone from *60 Minutes* to the president of the American Beekeeping Federation has been asking it," says entomologist Jeffery S. Pettis, research leader of the ARS Bee Research Laboratory in Beltsville, Maryland. "I wish the answer was as simple as the question."

Four broad classes of potential causes are being studied by ARS scientists and many others around the country and the world: pathogens; parasites; environmental stresses, which include pesticides; and management stresses, including nutrition problems, mainly from nectar or pollen dearth.

"What I believe is that CCD is likely a combination of factors, as opposed to a single, discrete cause," Pettis says. "When you do experimental studies, it's hard to isolate significant differences when you have more than one variable at a time," Pettis explains. Pettis has already planned several collaborations to look concurrently at two factors as possible causes. One will be a combination of exposure to pesticide and to the Israeli acute paralysis virus (IAPV), a virus shown to be strongly associated with CCD in a study—published in *Science*—that Pettis and colleague Jay D. Evans coauthored with university researchers.

The second experiment will look at the effect of a combination of varroa mites and pesticides. "If we find neither of these cause CCD, then we will go on to other combinations," Pettis adds. "And of course, there are other researchers around the world doing their own studies."

Another issue complicating the research is that, so far, researchers only have samples taken after a CCD incident is reported. With just the one set of samples, especially since the adult bees have disappeared, researchers cannot look for specific changes in affected bee colonies preceding the collapse.

To deal with this, in February 2007, Pettis and cooperators from universities and states began taking samples about every six weeks from three cooperating beekeepers in Florida that transport bees along the East Coast to provide pollination services. Two of the apiaries suffered outbreaks of CCD in 2006. If any of these apiaries have another outbreak of CCD, there will be samples that can be used to track changes over time, hopefully giving researchers a chance to see what changed, and potentially pinpointing the direction for future research.

IAPV and Other Pathogens

When IAPV was connected to CCD, researchers found IAPV in bees imported from China and Australia, which had only just begun shipping honey bees to the United States in 2005.

So entomologists Yanping (Judy) Chen and Evans, both with the ARS Bee Research Laboratory, conducted a detailed genetic screening of several hundred honey bees that had been collected between 2002 and 2007 from colonies in Maryland, Pennsylvania, California, and Israel. "Our study shows that, without question, IAPV has been in this country since at least 2002," said Chen. "This work makes it clear that IAPV is not a recent introduction from Australia."

But it also neither rules out nor reinforces the association between IAPV and CCD; it just settles the question of whether the recently imported Australian bees were the original source. The historical presence of IAPV in the United States does lead to a new question: Are there differences in virulence between imported and domestic IAPV strains? Chen is currently determining whether phenotypic differences exist among different viral strains. She'll then try to identify the genes responsible for differences in the virulence. To find out whether pathogens—IAPV or others—are indeed one of the factors in CCD, Evans has been collaborating with researchers from the University of Illinois to test honey bees from healthy and CCD-afflicted colonies as well as pre-CCD samples from 2002 on, for their abilities to mount immune responses.

In honey bees, exposure to pathogens activates the immune system, and different immuno-response genes are switched on by different pathogens. On the other hand, exposure to pesticides does not trigger immuno-response genes but arguably activates a different set of detoxifying genes. Evans and colleagues are looking for patterns in which either immune-related or detoxifying genes are activated in CCD versus healthy bees. "Once we've analyzed the data from this study, we hope we'll know whether pathogens or pesticides or both are factors to pursue," Evans said.

Some people believe that pesticides, especially a relatively new class called neonicotinoids, are responsible for CCD, though there is no conclusive data on this yet. France banned the neonicotinoid imidacloprid in 2005, when some field studies indicated some possible harm to bees, though other studies showed no such effects. And there has been no across-the-board recovery in honey bee populations in France since the ban.

The U.S. Environmental Protection Agency requires that companies provide data on a pesticide's possible impact on nontarget organisms before a pesticide can be registered for use, and honey bees are usually one of the nontarget insects tested. The neonicotinoids, which are based on nicotine, did not harm bees at the levels to which they are likely to be exposed.

But pesticide involvement in CCD is a possibility that is not being ruled out at this time. So Pettis and his colleagues are testing samples of bees, honey, wax, pollen, and nectar from CCD-afflicted and nonafflicted colonies for a wide variety of pesticides to see whether there are any patterns of pesticide residues that could contribute to CCD. "So far, we've found higher-than-expected levels of miticides that beekeepers use in the wax plus traces of a wide variety of agricultural chemicals in the pollen and wax, though there was no consistent pattern in either the levels or the types of chemicals identified," Pettis says. "No significant levels of agricultural chemicals were found in any honey." For now, researchers are casting a wide net to develop a science-based picture of what factors may result in CCD.

Sweet Temptations

In response to losses of honey bees, Häagen-Dazs, the maker of premium ice cream and sorbet, launched the Häagen-Dazs Loves Honeybees campaign. To date, the company has donated over $500,000 to University of California, Davis and Penn State for research to save the honey bee. A scientific advisory board was created, the educational Helpthehoneybees.com website was launched, and a new flavor was introduced: Vanilla Honeybee.

Chapter 5

Harvesting
The Honey

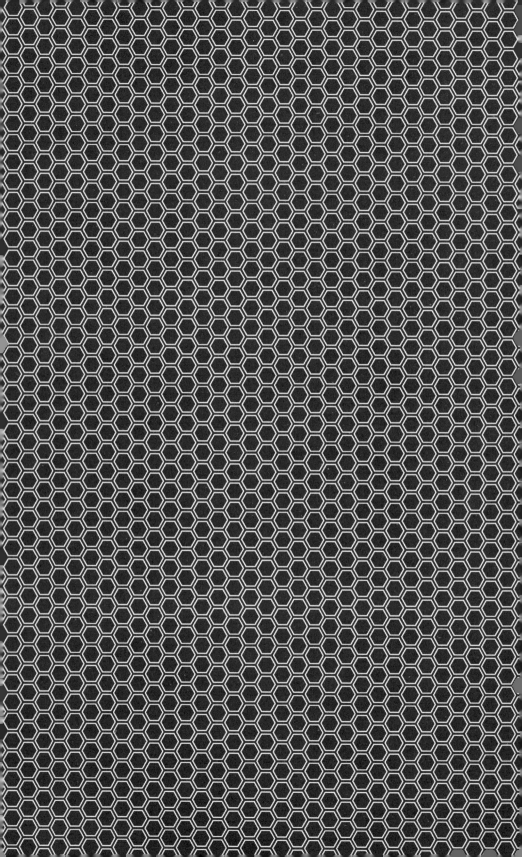

*L*et's face it: The hobby of beekeeping may be fascinating in and of itself, but the real allure of beekeeping is in the honey. Being able to put a jar of honey on the table that's made by your own bees is definitely a thrill, and something you should be rightfully proud of. If your hobby grows to the point that you can produce enough honey to give or sell to others, you'll experience the additional delight of making others' lives sweeter with your bees' honey.

This all sounds lovely—and in the end, it is—but getting from the combs to the packaged honey is a tricky, sticky process, one that should be well thought out ahead of time, as it has its own demands of equipment, time, and attention to detail.

CONSIDER YOUR OPTIONS

There are actually four styles of honey you can produce from your bees—extracted, comb, chunk, or whipped. Each has its own method of production that requires certain tools. Extracted honey is the type that is most familiar. To get it, the wax cappings must be sliced off and the liquid released through centrifugal force. Once extracted, this honey must be filtered before being bottled.

Comb honey is the honey in the comb. You can remove this from the frames and just eat it as is, wax and all. If you take chunks of the comb honey and put them in jars with extracted honey around them, it's called chunk honey or cut comb.

Whipped honey is extracted honey whose crystallization process has been controlled so that it forms a thick but smooth, spreadable product. Especially popular in Europe, whipped honey is also called creamed honey, spun honey, churned honey, or honey fondant. This type is the most labor intensive, but you may find that it yields a product that you and others especially love, which is different than the more typical extracted honey sold at the farmer's market (when or if you ever want to sell your honey).

Comb honey and whipped honey are specialized products that you can learn how to make by working with a beekeeper who knows how, or by researching the methods in other books, magazines, or online.

Advice for First-Timers

If you are a novice beekeeper, you can learn the most and have more fun by hooking up with fellow local beekeepers to observe how they extract and process their honey. If you think your hive has produced sufficient honey in its first year to yield a crop (this isn't always the case), you may want to borrow your fellow beekeepers' equipment so you can get a feel for it the first time out.

EQUIPMENT

To process honey, you'll need some basic supplies. For retailers of these items, see the Resources section at the back of this book.

Honey Extractor

This mechanism is what's used to spin the uncapped combs on the frames so that the honey is released by centrifugal force. You can find models in multiple sizes and with varying capacities, from small, hand-cranked extractors to giant ones operated on electric motors. For most hobby beekeepers, the smaller ones typically suffice—at least in the beginning. Even these should hold at least four frames at a time.

Before investing in this key piece of equipment, it's helpful to really investigate what's out there. There are even some clubs that rent the equipment, which is the best option for a beginner, though it's important to be sure the extractor is spotless and operational before you put your frames in it and operate it.

Uncapping Knife, Fork, and Tank

These are the tools that are used to strip the seals off of the cells in the comb that contain the honey. Because a heated and sharp edge is ideal for the job, today's uncapping knives are electric. Generations of beekeepers before you had to dip their long knives into buckets of hot water to keep them warm for the uncapping job. Be thankful for this handy tool, which also has serrated edges specially designed to uncap with care and precision. It's nice to have an uncapping fork handy, too. You may find you need something to remove stubborn caps on cells, and that's what the fork is designed to do.

An uncapping tank is a receptacle for catching the wax caps as they are dislodged from the comb. Certainly you can use something else—a large mixing bowl, for example—but if you think your hobby is going to grow, then an investment in an uncapping tank can make your life easier in the long run.

Bottling Bucket

Extracted honey is released from the extractor through a spigot at the bottom. When you're ready to drain the extractor, you'll need a large container for the honey. Bottling buckets are made from food-grade plastic, hold up to five pounds, and have a spigot at the bottom for ease of pouring honey into jars. Very handy!

With its handy spigot on the bottom, this five-gallon bucket is a perfect container for honey that's about to be put into jars.

Strainers, Filters, and Jars

Because the extracted honey will also have bits of gunk in it (wax, dirt, maybe a bee wing, etc.), it will need to be cleaned before bottling. A strainer or filter is necessary for this.

Jars are what you will put your honey into. Beekeeping supply catalogs have all sorts of jars and containers to choose from—everything from a basic jar to plastic containers with squirt tops to fancifully shaped bottles. For presentation purposes, the container you choose can make a statement. If you just want to bottle the honey for your own personal use, the jam or mayonnaise jars you have in the house should work fine.

WORKSPACE

Be forewarned: When you start working with the honeycomb-laden supers and the honey starts to flow, everything around you will get sticky—your equipment, your hands, your clothes, the countertops, the floor, the door knob—anything and everything in the room! Things that are sticky attract dust and dirt, which become even harder to get rid of because now they're dirty *and* sticky. Therefore, the room you select to work in must be as clean and neat as possible. Wash and dry every surface. Put plastic sheets or newspapers on the floor and plastic bags on the door handles. Be sure that bees can't make it into the space where you'll be working. They will be attracted to the honey, and as you will have learned by this stage, they can get into very tight spaces. Bee-proof your workspace.

Before doing anything with your supers, make sure all of your equipment is clean and that it is placed where you'll need it when you're ready to start processing the honey. A room that's suitable for the honey-making process is one that's easily cleaned and that is accessible to the outdoors. If you have an outdoor shed that's not too dirty, you can do it in there. A kitchen or mud room is another good choice—especially as these tend to have sources of hot and cold running water.

HARVESTING HONEY

Successful harvesting requires lots of planning and forethought. Besides the equipment and workspace considerations already discussed, you need to be sure you have enough time, and environmental conditions that are as conducive to a good harvest as possible. How much time you'll need depends on several things, most notably your experience. Even if you've worked with other beekeepers, expect the process to take longer if you're doing it by yourself for the first time.

Even experienced beekeepers run into unforeseen problems. They're unforeseen, of course, which makes factoring in time requirements difficult, so just know that no matter how long you think it might take, it'll take longer. For timing purposes, there are several steps involved. These are:

1. Prepare your workspace.

2. Remove the supers from the hive.

3. Extract and bottle the honey.

If you want to do this on a weekend, it would be best to spend Saturday morning or early afternoon getting your workspace cleaned and set up, Saturday afternoon or early evening removing the supers, and Sunday morning extracting and bottling the honey. Better to give yourself a couple of days to do it properly than to rush through the process and risk making costly mistakes.

When it comes to environmental considerations, you need to be your own judge. The best time to harvest is when a strong nectar flow is complete and you see that the hive is filled with tightly capped cells of honey. Depending on where you live and what's happening with Mother Nature in any particular season, you may have a very strong nectar flow in late spring that will yield enough honey for an early harvest. Whether or not you get one of those, you should have enough honey to warrant a harvest by the end of the summer—late August to October, depending on where you are. Remember, during the first year the hive will probably not yield enough for a harvest. Don't fret;

help a fellow local beekeeper so you can learn the intricacies of it, and look forward to next season.

First Things First: Retrieving the Supers

Get your protective clothing on, because when you check your supers, some bees will stay on them as you pick them up and move them around. This is fine for the occasional inspection, but when your supers are ready for extraction, all bees need to be off of them.

First, smoke the hive the way you normally would for an inspection. As you remove the frames from the super(s), use a bee brush to gently brush them off the frame. Brush in an upward direction so the comb isn't disturbed. You can also shake the frames to encourage the bees to leave. When the bees are off it, put the frame in the empty super and cover it with a board or towel, so bees can't get back to it. This is time consuming.

This short brush with super-soft bristles is another tool to use to gently and effectively encourage bees to leave the frames while you're harvesting honey.

You can also try installing a bee escape board between the brood hive and the super. It allows bees to travel down into the brood chamber easily, but makes it hard for them to get back into the super. It takes a few days to work, though, so it should be installed at least forty-eight hours before you want to remove the supers.

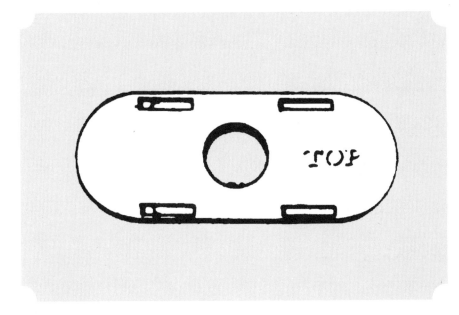

This small piece of equipment is placed in the hole of an inner cover and under a super you want to remove. Bees can leave the super, but they can't get back in.

Another option is to use a fume board and repellent. The fume board is a board lined with material to which a repellent has been applied. Placed on top of the supers (where the inner and outer covers would normally be), the scent causes the bees to move down into the brood chamber to get away from the smell. They move quickly, and you may have bee-free supers in as little as five minutes. Once they're in the brood chamber, remove the super, with speed and deliberation, to the honey harvesting area that you have bee-proofed. Because a full super can weigh more than thirty pounds, you'll want to have a hand truck or wheelbarrow handy. While the repellent used in the not-so-distant past was a chemical with some hazardous properties, today's repellent of choice is one made from natural ingredients and is nontoxic. Called Fischer's Bee Quick (see Beekeeping Resources), while its smell repels honey bees, its vanilla-almond scent is actually pleasant for humans. In an ideal world, supers would be removed, taken to the harvesting area, and processed shortly thereafter. Because removing the supers should be done when bees are calmest, typically after foraging time in the late afternoon or early evening, this isn't always possible.

Many beekeepers store the supers overnight and then extract in the morning. Remember that the longer the supers sit, the more danger there is of them being invaded by pests, noxious fumes, and other things that can spoil the honey. If you're going to store supers for several hours or overnight, be sure they're completely covered and are secure in your bee-proofed room where there is no chance that anyone will spill garbage or detergent on them or overturn them. Also, the room where you store the supers should be warm so the honey is easier to work with. A temperature of 70°F to 80°F is desirable.

Something else to know about honey is that it is hygroscopic, meaning that when it is exposed to air it will attract moisture from the atmosphere. In a very dry, warm atmosphere, the honey will lose water, and the quality will improve. In a cool, damp climate, it will absorb water from the air, which lessens the quality of the honey and can also initiate fermentation. Watery honey is not ready for extraction and should be returned to the hive. If you see honey running from open cells, bubbly honey, or honey weeping through cappings—even if it's just in a few cells—it's not ready.

Ready, Set, Extract!

Your equipment's ready, your supers are lined up. Hopefully you've had some experience either observing or working with another beekeeper, but now it's your moment of truth. Good luck!

Here's how to do the extraction, step by step:

1. Uncover the super you want to start working with, and then proceed frame by frame, keeping the other frames lightly covered with a towel while you work.

2. Position the frame over the uncapping tank or a large bowl, with it tipped slightly forward, and start to slice away the caps with your electric uncapping knife. Work from the bottom to the top using a side-to-side slicing motion. The wax caps will fall into the bowl and some honey will drip out but most of it will stay in the cells.

3. Use the uncapping fork to pry off any stubborn caps.

4. Turn the frame over and do the same thing to the other side.

5. Once the caps are removed, place the super into the extractor in a vertical position.

6. When you've filled the extractor with the amount of frames it can properly hold, start cranking! If yours is a hand-cranked extractor, start slowly and don't build up too much speed. You don't want the wax comb to be damaged, only the honey removed.

7. When cranking becomes difficult, you'll know the extractor is filling with honey. Stop cranking, position a bucket and honey strainer under the nozzle, and strain the honey into the bucket.

8. Once all the honey is extracted and strained, line up your jars and begin filling them. Secure them with tight-fitting lids, wipe them down to remove any stickiness, place them on a shelf away from direct sunlight, and get ready to clean the extractor.

Managing the Extractor

A filled extractor can get heavy and, like a washing machine when a heavy load is unevenly distributed, can start to shake and wobble as honey is extracted from the frames. Working a manual extractor with an unbalanced load is fatiguing; on an electric-powered one, it can be dangerous.

To manage the imbalance, observe the frames as they're placed in the extractor, loading them so that the heavy and light frames are evenly distributed. If any granulation (separation) occurs in the frames, the imbalance may increase as the liquid honey is thrown out and the solid remains. Wherever possible, extractors should be secured to an immovable surface, though this is often impractical. One solution for a small manual extractor may be to mount it on a board that extends to the operator's position, so his or her weight is anchoring the machine. Be sure the extractor is high enough so the honey can flow directly through a honey strainer and into a bucket.

Granulated Honey

An extractor cannot remove granulated honey, and there is no way to practically remove it without destroying the honeycomb. Rather than overwork the honeycomb, simply place the supers with granulated sugar in the combs back into the hive box in the early spring. The bees will love it, and they will get in and clean the combs while they consume the granules. Scratch the caps with your uncapping fork before you put the supers back so that the bees can access the cells.

Finishing the Job

Just as you started with clean equipment, you need to clean and store your equipment and supplies so they're ready for the next harvest. The number of times you harvest during a season will depend on how many hives you keep and how productive your bees are. It's typically once or twice a season (early and late summer). Cleanup is relatively simple: Knives, strainers, buckets, and so on should be washed in warm soapy water, rinsed thoroughly in warm/hot water, and stored away.

To clean the extractor, first rinse with cold water so that the waxy bits don't melt into the machinery. Continue rinsing with successively warmer water to continue to flush away dirt and particles. The last rinse will be the hottest, but it shouldn't be scalding, as you don't want wax to streak the surfaces and get stuck on them. Use a long-handled brush to loosen debris, and when thoroughly cleaned, use a soft absorbent cloth to dry everything as well as you can before you put it away.

YOUR WAX HARVEST

You'll need to do something with the wax and debris that came off when you uncapped the combs. It's a shame to waste these wonderful by-products. The wax can be processed and used to make all kinds of things, and the residual honey can be fed back to your bees or used to make mead. Howland Blackiston, author of *Beekeeping for Dummies*, follows these five steps for getting the most out of the cappings.

1. Allow gravity to drain as much honey from the cappings as possible. Let them sit for a few days, either in your uncapping tank or wrapped in cheesecloth and suspended over a bowl.

2. Place the drained cappings in a five-gallon plastic pail and top them off with warm (not hot) water. Then, using a large spoon or even your hands, slosh the cappings in the water to remove any remaining honey. Drain the cappings through a colander or honey strainer and repeat as many times as necessary for the water to run clear.

3. Place the washed cappings in a double boiler and melt the wax. Never melt wax directly on a heat source, and don't let it sit unsupervised for any amount of time.

4. Strain the melted beeswax through a couple of layers of cheesecloth to remove any debris.

5. If you're not ready to use the wax right away, pour it into a block mold for later use. Blackiston uses a clean cardboard milk carton that will hold the block until it's ready to be used, at which time the carton can be easily torn to reveal the wax.

Chapter 6

Using
The Honey
And Beeswax

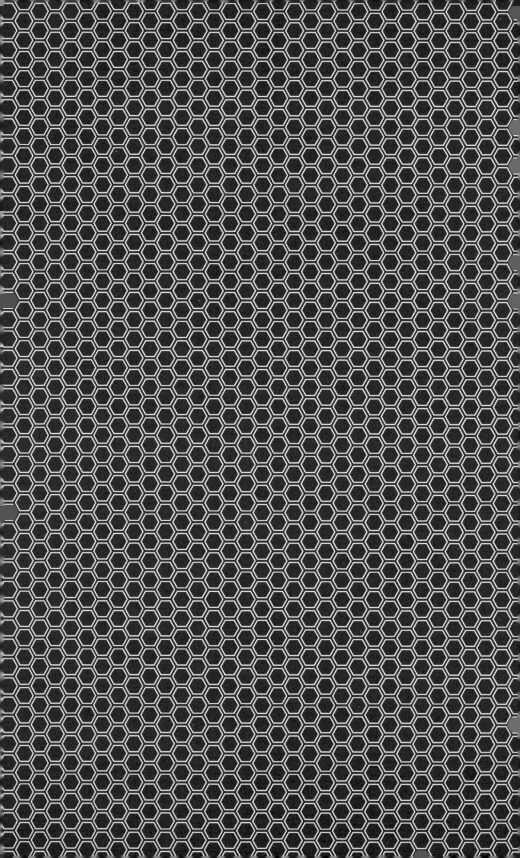

*J*ust the sight of honey elicits a mouth-watering response, and it is easy to understand why it is called "liquid gold." Honey was gold to the people who were able to collect and process it, especially when it was the only sweetener available. Besides its key role as a sweetener, honey has also been used through the ages for its healing powers and for its beauty-enhancing properties. But wait, there's more!

An additional useful and valued by-product of the honey bees' activities is beeswax, which is an ingredient— sometimes the main ingredient—in fine products such as candles, health and beauty aids, household products, and more. This chapter explores honey and beeswax in their many forms and uses.

Many of us are familiar with the honey found in our local grocery stores, but those mass-produced bottles and jars contain only a small portion of the types of honey that can be found worldwide. In the United States alone, there are more than three hundred varieties of honey, differing in flavor, fragrance, and color, depending upon the plants the bees visit and the geographic area. In general, lighter-colored honeys tend to be milder, whereas darker-colored honeys have stronger flavors. Let's take a look at some of the more popular varieties, based on the kinds of plants that produce them.

VARIETIES *of* HONEY

Alfalfa honey, flavored by the legume's blue flowers, is white or extra light amber in color with a fine flavor. The honey has good body, which makes it a perfect table honey. Alfalfa blooms throughout the summer and is ranked as the most important honey plant in Utah, Nevada, Idaho, Oregon, and most of the western United States.

Avocado honey is gathered from California avocado blossoms. Avocado honey is dark in color, with a rich, buttery taste.

Basswood is a tree distributed from southern Canada to Texas. Basswood honey is often characterized by its distinctive bite of flavor.

Sourwood

Buckwheat

Blueberry

Blackberry Tree

Clover

Tulpo Tree

The flowers are cream colored and bloom in late June and July. The honey is "water white" in color with a strong flavor.

Blueberry nectar is taken from the tiny white flowers of the blueberry bush, and makes a honey that is typically light amber in color and with a full, well-rounded flavor. Blueberry honey is produced in New England and in Michigan.

Buckwheat plants grow best in cool, moist climates. Buckwheat prefers light, well-drained soils, although it can thrive in highly acidic, low fertility soils as well. Buckwheat is usually planted in the spring, blooms quite early, and yields a dark brown honey of strong, distinct flavor.

Clovers are the most popular honey plants in the United States. White clover, alsike clover, and the white and yellow sweet clover plants are the most important for honey production. Depending on location and source, clover honey varies in color from water white to extra light amber and has a mild, delicate flavor.

Eucalyptus is one of the larger plant genera with more than five hundred distinct species, and many hybrids. Eucalyptus honey varies greatly in color and flavor, but in general, it tends to be a bold-flavored honey with a slightly medicinal aftertaste.

Fireweed honey is light in color and comes from a perennial herb that affords wonderful bee pasture in the northern and Pacific states and Canada. Fireweed grows in the open woods, reaching a height of three to five feet, and spikes attractive pinkish flowers.

Orange blossom honey is often a combination of citrus floral sources. Orange is a leading honey source in southern Florida, Texas, Arizona, and California. Orange trees bloom in March and April and produce a white to extra light amber honey with a distinctive flavor and the aroma of orange blossoms.

Sage honey can come from different species of the plant. Sage shrubs usually grow along the California coast and in the Sierra Nevada. Sage honey has a mild, delicate flavor. It is generally white or water white in color.

Sourwood trees can be found in the Appalachian Mountains from southern Pennsylvania to northern Georgia. Sourwood honey has a sweet, spicy anise aroma and flavor with a pleasant lingering aftertaste.

The **tulip poplar** is a tall tree with large greenish yellow flowers. It generally blooms in the month of May. Tulip poplar honey is produced from southern New England to southern Michigan and south to the Gulf states east of the Mississippi. The honey is dark amber in color; however, its flavor is not as strong as one would expect from a dark honey.

Tupelo honey is produced in the southeastern United States. Tupelo trees have clusters of greenish flowers, which later develop into soft, berrylike fruits. In southern Georgia and northwestern Florida, tupelo is a leading honey plant, producing tons of white or extra light amber honey in April and May. The honey has a mild, pleasant flavor and will not granulate.

HONEY *for* HEALTH *and* BEAUTY

Honey is good for you in so many ways! Our ancestors knew this, and we are able to both learn what they knew about honey and expand on it ourselves. That the honey bee is responsible for the full ripening of so many of our foods makes it eminently significant to our survival as a species.

What makes honey such a health benefit (besides a tasty treat) is the fact that it contains a number of vitamins and minerals. Niacin, riboflavin, and pantothenic acid are all important B vitamins that occur naturally in honey. The minerals it contains include calcium, copper, iron, magnesium, manganese, phosphorous, potassium, and zinc—a nutritional bonanza!

Depending on the type, honey also has significant antioxidant potential; darker honeys are believed to have higher antioxidant levels. Add to that one of the best sources of pure carbohydrate (17 grams per tablespoon) and a healthy dose of water, and you can understand how honey has played such a healing role through time.

Medicinal Qualities

Honey and other products associated with bees, including pollen, royal jelly, and even bee venom, have all been used and studied to treat a variety of health disorders. Apitherapy is the term applied to the use of bees or bee products to heal or recover someone from an illness.

Apitherapy as an already large and growing body of medicine continues to produce surprising results, and it cannot be adequately covered in this book. It is fascinating, though, and certainly worth exploring, that apitherapy can help treat a range of ailments from common coughs to chronic sciatica and even cancer.

A "short list" of ailments that have been documented to respond to apitherapy includes acne, AIDS, allergies, anemia, arthritis, bronchitis, ear infection, eczema, gingivitis, hair loss, hay fever, heart disease, high blood pressure, hepatitis, lumbago, mononucleosis, multiple sclerosis, PMS, sexual dysfunction, tuberculosis, warts, and wounds. (See Beekeeping Resources section for more information.)

Honey Helping Bones

Calcium is a mineral that helps combat the development of osteoporosis (low bone mass). To do its job right, it has to be adequately absorbed into the bloodstream, and it turns out that honey can help its absorption rate. A Purdue University study published in 2005 found that calcium absorption in lab animals was definitely enhanced when honey was consumed. The study even showed that the more honey consumed, the higher the rate of calcium absorption. While the results may vary with human beings, it is definitely compelling—especially for athletes and older people.

Honey for Beauty . . . and Burt

Now that we've explored how honey and bee products are good for you on the inside, it's fun to learn some ways that honey and bee products are good for you on the outside as well. Honey-based beauty aids go back to ancient times, and it is well known that Cleopatra was a big fan of honey as a way to keep her skin moist and her hair shiny. Today we have a huge assortment of honey and bee beauty products available to us. Honey is nonirritating, which is great for sensitive skin, and it has antibacterial properties that promote healing. One of the largest suppliers of these products in the United States is Burt's

Bees—and the company's story is an inspiration to all beekeepers who aspire to do more with what their bees make.

Burt's Bees started in 1984 with Roxanne Quimby and Burt Shavitz in Maine. Roxanne used Burt's beeswax to make candles and started selling them at craft fairs, where they became a thriving business. As demand grew, production increased and creativity flourished. After reading about ancient beauty and skincare products made from bees, Roxanne started experimenting. Just ten years after the first candles were fashioned, Burt's Bees moved to North Carolina, where an eighteen-thousand-square-foot facility became the headquarters for a product line that had grown to fifty items worldwide. Today the company features more than 150 items, and its sales are well over $250 million a year. Its growth is a testament to the quality of the products, the appeal of natural bee-based items, and the embrace of earth-friendly packaging, processes, and philosophies. Could you be the founder of the next Burt's Bees?

One of the things that makes honey so great for skin is that it is a natural humectant—it naturally attracts and retains moisture. This is a concern for beekeepers, as the honey that sits in a warm, moist area for too long weakens the cells of the super.

The tricky thing about honey is that it's sticky—and has a sweet smell that attracts ants, flies, and other insects and animals. Health and beauty product companies spend time and money figuring out how to minimize the stickiness, to our benefit, but there are a number of home remedies that you can try. Just don't plan to go anywhere until you've been able to rinse them off!

HOMEMADE HONEY TREATMENTS

The following treatments are courtesy of the National Honey Board.

⋯⋯ HONEY CLEANSING SCRUB ⋯⋯

Mix 1 tablespoon honey with 2 tablespoons finely ground almonds and ½ teaspoon lemon juice. Rub gently onto face. Rinse off with warm water.

⋯⋯ FIRMING FACE MASK ⋯⋯

Whisk together 1 tablespoon honey, 1 egg white, 1 teaspoon glycerin (available at pharmacies and beauty supply stores) and enough flour to form a paste (approximately ¼ cup). Smooth over face and throat. Leave on 10 minutes. Rinse off with warm water.

⋯⋯ SMOOTHING SKIN LOTION ⋯⋯

Mix 1 teaspoon honey with 1 teaspoon vegetable oil and ¼ teaspoon lemon juice. Rub into hands, elbows, heels, and anywhere that feels dry. Leave on 10 minutes. Rinse off with water.

⋯⋯ SKIN SOFTENING BATH ⋯⋯

Add ¼ cup honey to bath water for a fragrant, silky bath.

⋯⋯ HAIR SHINE ⋯⋯

Stir 1 teaspoon honey into 4 cups (1 quart) warm water. Blondes may wish to add a squeeze of lemon. After shampooing, pour mixture through hair. Do not rinse out. Dry as normal.

WHAT *to* DO *with* BEESWAX

After you've extracted honey from your supers, you will be left with the wax cappings and other detritus from the combs. The process for cleaning this so that you are left with the pure beeswax was described in Chapter 5. Once cleaned and poured into blocks, the wax can be worked to create any number of things but most typically candles, cosmetics, and furniture polish. Creating handcrafted products from your beeswax is another way to showcase the wonder of your bees, and to delight family and friends.

Candles

There are several materials that form the basis for most of the candles we use today: paraffin, soy, and beeswax are among them. The paraffin wax used for candles is a petroleum derivative, and will drip, smoke, and sputter as the wick absorbs and burns it. Soy wax is created from soybean oil, and is a natural wax form that is becoming more and more popular. But beeswax is the ultimate wax when it comes to candles. Beeswax doesn't smoke, sputter, or drip, it has a lovely sheen and texture, and it can be crafted from sheets into rolled candles, molded into tapers, or hand dipped in various sizes.

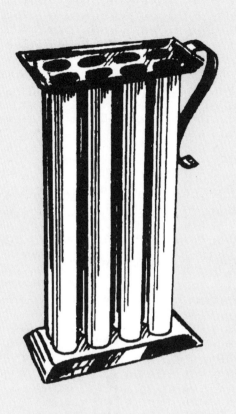

One way to make consistent-sized candles is to use a mold, like this taper mold, which makes eight candles.

Candle making starts with melted wax. Remove some of your cooled blocks from the milk cartons or containers you've kept them in, and put them in the top of a double boiler. Don't heat or melt wax directly over a heat source. Like chocolate, you want it to come to the proper liquid temperature slowly and steadily so that it retains its consistency and color.

Rolled candles are made by pouring melted wax onto a rectangular or square surface—a nonstick cookie sheet or brownie pan, for example. When the wax has cooled enough to handle it (but not so cool that it will crack and break), the sheet can be removed onto another nonstick surface, a wick positioned on one end, and the sheet carefully rolled into shape. Trim the wick, melt and smooth the bottom so it is flat, and allow to cool thoroughly.

Molded candles are just that: candles made from wax that's been poured into molds. These are available in a large variety of shapes and sizes and require nothing more than properly positioned wicks in the molds, the melted wax poured into them and then allowed to cool.

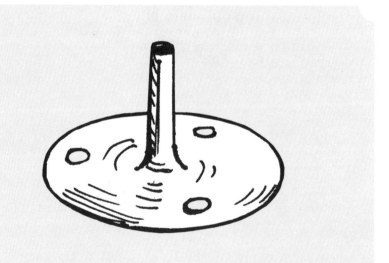

One way to keep the wick straight is to use a T-Tab like this one.

Dipped candles are the most time consuming to make, but are the most beautiful and special when given as gifts. With a weighted wick to keep the candle straight as it is formed, the candlemaker simply dips the wick repeatedly into melted wax, allowing each dip to cool sufficiently before dipping again.

What Is Beeswax?

Beeswax is made by worker bees when they build the combs that house the bee brood and bee food (honey!). It is pale when it is first made, but becomes more discolored as it ages and is tarnished with pollen and propolis. Beeswax is made up primarily of hydrocarbons, monoesters, diesters, and free acids. It has a very high melting point, and is quite resilient.

Cosmetics

Making things like lip balm or skin cream with beeswax begins in the same way as candle making: with the block of wax you cleaned and stored after extracting honey from the supers. Just as honey is a natural medicinal treatment because of the way it reacts with and retains water, so beeswax is a wonderful addition to skin treatments as it, too, assists in moisture retention and also smooths and lubricates dry skin—especially when paired with other beneficial oils and essential oils for fragrance.

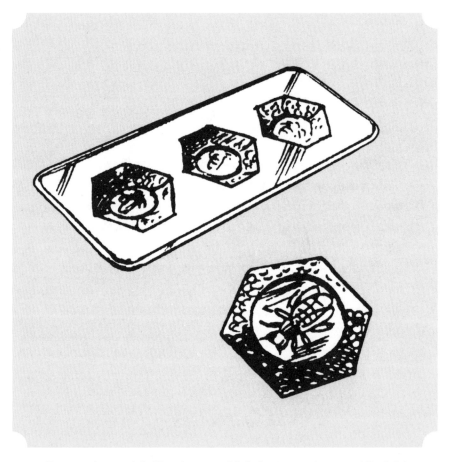

Decorative molds like these, which feature a bee motif, yield beautiful pieces of soap.

If you are interested in crafting candles or cosmetics, there are many resources to help you further your creativity and refine your products. There are simple things like adding colors, scents, and patterns to the wax for candles or soaps. And there are complicated things like making stand-out formulations for skincare products.

Explore your options through books, websites, and, of course, fellow crafters. Check out the Resources section of this book for some ideas.

Furniture Polish

Beeswax makes an excellent furniture polish when combined with a few other key ingredients. The following recipe, from Care2.com, is a beeswax furniture polish that is simple, earth friendly, and consistent with other formulations. Have a container at the ready for the polish when it's prepared. Glass baby food jars work very well. Keep the polish tightly sealed in a cool, dark area for maximum shelf life.

Basic Polishing Cream Waxing Formula

4 ounces oil (2½ ounces olive oil or jojoba,
   1½ ounces coconut oil)
1 ounce beeswax
1 ounce carnauba wax
4 ounces distilled water

Melt the oils and waxes in a double boiler over medium heat. Remove from the heat, pour in the water, and mix with a hand mixer until thick and creamy.

Dab some cream onto a soft cotton rag and rub into the furniture. Buff and polish until the oils are well worked into the wood.

Shelf life: 6 months to a year.

Historical Uses of Beeswax

Its strength and resistance to heat made beeswax the molding material of choice for our ancestors in sculpture and jewelry (using a method called the lost-wax casting process). The ancient Egyptians used beeswax in pursuits as varied as shipbuilding and mummification. Artists, from cave painters to Renaissance artists, preserved their pigments and finished works, and the makers of many kinds of musical instruments used beeswax for everything from mouthpieces to bows. And of course it was used in many medicinal and cosmetic preparations then, as it is today.

Chapter 7

Cooking
With Honey

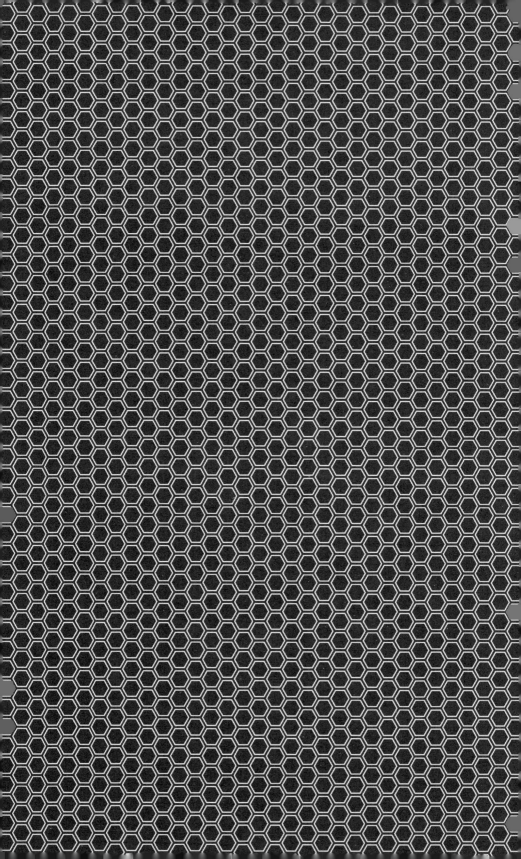

W hat's your favorite way to serve or cook with honey? Is it something as simple as spreading honey over a piece of warm buttered toast, or maybe swirling it over granola and yogurt, or ice cream, or a fruit dessert? While we typically think of honey as a simple and satisfying sweetener, it is also a wonderful complement to savory dishes.

True honey conoisseurs may even prefer the honeys produced at different times of the year or from different flowers and plants. As you produce honey from your bees, you will soon become sensitive to—and appreciative of—the variations in flavors. Whether you choose to showcase these in recipes that are all about the honey or use them to complement other foods, it'll be fun to experiment and receive reactions from your family and friends.

COOKING WITH HONEY

If you're like a lot of people, you may have the desire to substitute processed sugar with honey in your diet. The following recipes are from the website of the National Honey Board, the organization committed to promoting honey in the United States. This is just a smattering of the recipes available on their website at Honey.com, where you can sign up for a newsletter to receive meal plans that feature honey, often with recipes developed by top-notch chefs. Dig in!

The recipes are categorized as follows:

- Drinks and Snacks Recipes
- Breakfast and Smoothie Recipes
- Lunch and Dinner Recipes
- Side Dishes
- Dessert Recipes

Drinks and Snacks Recipes

HONEY HYDRATOR

$1/2$ cup honey
$1/2$ teaspoon lite salt
2 cups orange juice
$5 1/2$ cups lukewarm water

Combine ingredients, then cool before drinking.

FRUITY HONEY SMOOTHIE

1 cup frozen strawberries
1 banana
$1/3$ cup honey
1 cup nonfat milk
$1/2$ cup plain or vanilla low-fat yogurt

In a blender, combine all ingredients; process until smooth.

NO-BAKE HONEY ENERGY BARS

$^2/_3$ cup honey
$^3/_4$ cup creamy or chunky peanut butter
4 cups granola mix

In 4-cup microwave-safe container, heat honey on high for 2 to 3 minutes or until honey boils. Stir in peanut butter; mix until thoroughly blended. Place granola in large bowl. Pour honey mixture over granola and combine thoroughly. Press firmly into 13 x 9 x 2 inch baking pan lined with wax paper. Let stand until firm. Cut into bars.

Breakfast and Smoothie Recipes

EASY HONEY MUFFINS

$^1/_2$ cup milk

$^1/_4$ cup honey

1 egg, beaten

$2^1/_2$ cups buttermilk baking mix

Preheat oven to 400°F. Combine milk, honey, and egg; mix well. Add baking mix and stir only until moistened. Portion into greased muffin tins. Bake for 18 to 20 minutes or until wooden pick inserted near center comes out clean. Variation, Cinnamon Apple Muffins: Add 2 cups pared, chopped apples and 1 teaspoon ground cinnamon to basic recipe. Bake about 5 minutes longer than basic recipe.

COLUSA CORN MUFFINS

$^2/_3$ cup milk

$^1/_3$ cup butter or margarine, melted

$^1/_2$ cup honey

2 eggs

$1^1/_2$ cups whole wheat flour

$^2/_3$ cup cornmeal

$2^1/_2$ teaspoons baking powder

$^1/_4$ teaspoon salt

Preheat oven to 350°F. In small bowl, beat together milk, butter, honey, and eggs. Set aside. In large bowl, stir together dry ingredients. Add honey mixture. Stir just enough to barely moisten flour. Do not overmix. Spoon batter into paper-lined or greased muffin pan cups. Bake for 20 to 25 minutes. Serve warm.

FAT-FREE HONEY BERRY MILKSHAKE

1 pint nonfat vanilla ice cream or nonfat frozen yogurt
1 basket strawberries, hulled, or an assortment of berries,
  approx. 2½ cups
¹/₂ cup nonfat milk
¹/₄ cup honey
4 small mint sprigs, optional garnish

In blender, combine all ingredients except mint sprigs and blend until smooth and creamy, about 30 seconds. Pour immediately in tall, chilled glasses and garnish with mint sprigs.

GINGER PEACH SMOOTHIE

1 cup boiling water
1 piece (about 2 inches) fresh ginger root, peeled and
   crushed
$^1/_4$ cup honey
2 peaches, peeled, pitted, and chopped
1 pint peach sorbet
1 tablespoon lime juice

In small bowl, combine water and ginger. Stir in honey; cool. Remove and discard ginger. Set mixture aside. In blender or food processor container, combine peaches, sorbet, and lime juice. Process while adding honey-ginger mixture in a slow, steady stream; process until smooth.

Lunch and Dinner Recipes

BROILED LEMON HONEY CHICKEN BREASTS

$^1/_4$ cup honey

$^1/_4$ cup lemon juice

2 teaspoons vegetable oil

1 teaspoon rosemary, crushed

1 teaspoon grated lemon peel

$^1/_2$ teaspoon salt

$^1/_4$ teaspoon pepper

4 boneless, skinless chicken breasts, $3^1/_2$ to 4 ounces each

Combine all ingredients (except chicken) and mix well. Marinate chicken in honey-lemon mixture 1 hour in shallow baking dish. Broil chicken 5 minutes, brushing with pan drippings, turn and broil 5 minutes longer or until juices run clear. If desired, bring marinade to a boil; simmer 2 minutes. Strain hot marinade over chicken.

BANGKOK WRAP

1 cup grated carrots

1 cup grated cucumber

2 cups shredded cabbage

2 cups cooked rice

3 cups cooked, shredded pork roast

salt and pepper, to taste

$^{1}/_{2}$ cup honey

$^{1}/_{2}$ cup rice wine vinegar

2 tablespoons chunky peanut butter

2 tablespoons chopped fresh cilantro

2 teaspoons grated fresh ginger

2 cloves garlic, minced

8 large lettuce leaves (iceberg or green leaf), blanched
  and patted dry with paper towel

In medium bowl, combine carrot, cucumber, cabbage, rice, and pork. Season with salt and pepper.

In small bowl, whisk together honey, vinegar, peanut butter, cilantro, ginger, and garlic. Heat dressing in small saucepan, stirring until thickened, about 3 to 4 minutes.

To assemble wraps, lay lettuce leaves on work surface. Spoon ½ cup pork mixture in center of each leaf. Drizzle each wrap with 2 tablespoons dressing. Fold in sides to wrap and serve.

A HONEY OF A CHILI

1 package (15 ounces) firm tofu

1 tablespoon vegetable oil

1 cup chopped onion

$^3/_4$ cup chopped green bell pepper

2 cloves garlic, finely chopped

2 tablespoons chili powder

1 teaspoon ground cumin

1 teaspoon salt

$^1/_2$ teaspoon dried oregano

$^1/_2$ teaspoon crushed red pepper flakes

1 can (28 ounces) diced tomatoes, undrained

1 can (15½ ounces) red kidney beans, undrained

1 can (8 ounces) tomato sauce

$^1/_4$ cup honey

2 tablespoons red wine vinegar

Using a cheese grater, shred tofu and freeze in zippered bag or airtight container. Thaw tofu; place in a strainer and press out excess liquid. In large saucepan or dutch oven, heat oil over medium-high heat until hot; cook and stir in onion, green pepper, and garlic 3 to 5 minutes or until vegetables are tender and begin to brown. Stir in chili powder, cumin, salt, oregano, and crushed red pepper. Stir in tofu; cook and stir 1 minute. Stir in diced tomatoes, kidney beans, tomato sauce, honey, and vinegar. Bring to a boil; reduce heat and simmer, uncovered, 15 to 20 minutes, stirring occasionally.

Side Dishes

BEES IN THE GARDEN COLESLAW

1 head green cabbage, shredded
1 green bell pepper, diced
$1/2$ cup sweet red pepper, diced
$1/2$ cup mayonnaise
$1/3$ cup honey
2 tablespoons vinegar
$1/2$ teaspoon dry salt
$1/2$ teaspoon dry mustard
$1/2$ teaspoon celery seed
$1/4$ teaspoon black pepper

Toss cabbage and peppers in large bowl. Combine mayonnaise, honey, vinegar, salt, mustard, celery seed, and black pepper in medium bowl and then toss with cabbage mixture. Mix well. Cover and refrigerate until thoroughly chilled.

GRILLED CORN WITH
SPICED HONEY BUTTER

$^1/_2$ cup butter, softened

$^1/_3$ cup honey

1 teaspoon chili powder

8 ears fresh corn

8 lime wedges

$^1/_4$ cup fresh cilantro, chopped

In a small bowl, stir together the butter, honey, and chili powder; set aside. Fold back husks and remove silk from corn; pull husks back up over corn. Place corn in a large bowl of ice water and soak for 15 minutes. Remove and shake off excess water. Place on grill over medium-hot coals and cook for 15 to 20 minutes, turning frequently. Remove husks and spread each ear with seasoned butter. Sprinkle with cilantro and serve with lime wedges.

Dessert Recipes

BITTERSWEET CHOCOLATE RASPBERRY TRUFFLE CUPCAKES

CUPCAKES

8 ounces 60-percent-cocoa bittersweet chocolate, divided

2 cups unbleached all-purpose flour

$^1/_2$ teaspoon baking soda

$^1/_2$ teaspoon baking powder

$^1/_2$ teaspoon salt

$^1/_2$ cup (1 stick) butter, softened

$^3/_4$ cup clover honey

2 large eggs

1 cup buttermilk

$^1/_2$ pint raspberries

FROSTING

$^1/_4$ cup clover honey

8 ounces 60-percent-cocoa bittersweet chocolate,
   coarsely chopped

1 cup heavy whipping cream

2 tablespoons seedless raspberry jam, optional

Preheat oven to 350°F. Cut 2 ounces of chocolate into 18 pieces; set aside. Place 6 ounces of coarsely chopped chocolate in a microwave-safe dish. Microwave 30 seconds on high; stir well. Microwave 30 seconds more on high and stir until all lumps are gone. If more melting is necessary, microwave in 10 second increments and stir until all lumps are gone. Chocolate should not become too warm. Sift together flour, baking soda, baking powder, and salt; set aside. In a mixing bowl, cream butter until fluffy. Add honey and melted chocolate; mix well. Add eggs, one at a time. Add half of the dry ingredients to the butter mixture; mix on low until just combined. With mixer running on low, slowly add the buttermilk. Add remaining dry ingredients until just combined. Place a tablespoon of batter in each paper-lined cup in muffin tin. Add one piece of chocolate and 2 to 3 raspberries to each cup; fill each two-thirds full with remaining batter. Bake 18 to 22 minutes, or until toothpick comes out clean. Remove to wire rack; cool.

Frosting: Combine honey and chocolate in a medium bowl; set aside. In small, heavy pan, heat whipping cream over medium heat until bubbles just begin to form. Pour over honey-chocolate mixture and allow to stand for 2 minutes. Stir until smooth; cool. Refrigerate until chilled, 1 to 2 hours. With an electric mixer, beat chocolate mixture until frosting is fluffy.

CHOCOLATE ZUCCHINI CAKE

$^1/_2$ cup butter

$^1/_4$ cup vegetable oil

$^2/_3$ cup honey

2 eggs

1 teaspoon vanilla

$^1/_3$ cup buttermilk

$2^1/_2$ cups flour

4 tablespoons cocoa

$^1/_2$ teaspoon baking powder

1 teaspoon baking soda

$^1/_2$ teaspoon cinnamon

$^1/_2$ teaspoon cloves

2 cups grated zucchini

1 cup nuts, optional

Preheat oven to 350°F. Grease and flour a 13 x 9 x 2-inch pan; set aside. In the bowl of an electric mixer, beat together butter, oil, and honey. Add eggs, vanilla and buttermilk; mix well. Combine dry ingredients and add to honey mixture; mix just until combined. Stir in zucchini and nuts, if desired. Pour into prepared pan. Bake 30 minutes. Let cool to room temperature, then dust with powdered sugar, if desired.

Beekeeping
Resources

BOOKS

Benjamin, Alison, and Brian McCallum. *Keeping Bees and Making Honey* (Devon, UK: David & Charles, 2008).

Bishop, Holley. *Robbing the Bees: A Biography of Honey—The Sweet Liquid Gold that Seduced the World* (New York: Free Press, 2006).

Blackiston, Howland. *Beekeeping for Dummies* (Hoboken, NJ: John Wiley & Sons, 2002).

Bonney, Richard E. *Hive Management: A Seasonal Guide for Beekeepers* (North Adams, MA: Storey Publishing, 1991).

— *Beekeeping: A Practical Guide* (North Adams, MA: Storey Publishing, 1993).

Connor, Lawrence John. *Increase Essentials* (Kalamazoo, MI: Wicwas Press, 2006).

Conrad, Ross. *Natural Beekeeping: Organic Approaches to Modern Apiculture* (White River Junction, VT: Chelsea Green Publishing, 2007).

Crane, Eva. *The World History of Beekeeping and Honey Hunting* (New York: Routledge, 1999).

Dadant, C. P. *First Lessons in Beekeeping* (Eastford, CT: Martino Publishing, 2009).

Delaplane, Keith S. *Honeybees and Beekeeping: A Year in the Life of an Apiary*, 3rd ed. (Athens, GA: University of Georgia Center for Continuing Education, 2006).

Ellis, Hattie. *Sweetness and Light: The Mysterious History of the Honeybee* (New York: Three Rivers Press, 2006).

Flottum, Kim. *The Backyard Beekeeper: An Absolute Beginner's Guide to Keeping Bees in Your Yard and Garden* (Beverly, MA: Quarry Books, 2005).

Horn, Tammy. *Bees in America: How the Honeybee Shaped a Nation* (Lexington, KY: University Press of Kentucky, 2006).

Hubbell, Sue. *A Book of Bees: And How to Keep Them* (New York: Mariner Books, 1998).

Kidd, Sue Monk. *The Secret Life of Bees* (New York: Penguin, 2008).

Langstroth, L. L. *Langstroth's Hive and the Honey-Bee: The Classic Beekeeper's Manual* (New York: Dover, 2004).

Longgood, William, and Pamela Johnson. *Queen Must Die and Other Affairs of Bees and Men* (New York: W. W. Norton, 1988).

Ransome, Hilda M. *The Sacred Bee in Ancient Times and Folklore* (New York: Dover, 2004).

Root, A. I., and E. R. Root. *ABC and XYZ of Bee Culture* (Whitefish, MT: Kessinger Publishing, 1947).

Steiner, Rudolph. *Bees* (Herndon, VA: Steiner Books, 1998).

Tompkins, Enoch H., and Roger M. Griffith. *Practical Beekeeping* (Burlington, VT: Garden Way, 1977).

Vivian, John. *Keeping Bees* (Nashville, TN: Williamson Publishing, 1986).

MAGAZINES

American Bee Journal
Dadant & Sons, Inc.
51 S. Second St.
Hamilton, IL 62341
212-847-3324
www.dadant.com/journal

Bee Culture
A. I. Root Company
PO Box 706
Medina, OH 44256
800-289-7668
www.beeculture.com

Beekeepers Quarterly
"Britain's Biggest and Brightest Beekeeping Magazine"
Northern Bee Books
Mytholmroyd
Hebden Bridge HX7 5JS
United Kingdom
www.beedata.com/bbq.htm

Bee World
International Bee Research Foundation (IBRA)
16 North Road
Cardiff CF1 3DY
United Kingdom
www.ibra.org.uk

BEEKEEPING SUPPLIES

Bee-Commerce
11 Lilac Lane
Weston, CT 06883
800-784-1911
www.bee-commerce.com

Betterbee
8 Meader Road
Greenwich, NY 12834
800-632-3379
www.betterbee.com

Brushy Mountain Bee Farm
610 Bethany Church Road
Moravian Falls, NC 28654
800-233-7929
www.brushymountainbeefarm.com

All Equipment illustrations in this book were graciously provided by Brushy Mountain Bee Farm. Equipment shown as well as additional supplies for beekeeping can be purchased directly through their website or by calling.

Dadant & Sons, Inc.
51 S. Second
Hamilton, IL 62341
888-922-1293
www.dadant.com

Mann Lake Ltd.
501 S. First St.
Hackensack, MN 56452
800-880-7694
www.mannlakeltd.com

BEEKEEPING ORGANIZATIONS

American Apitherapy Society (AAS)
500 Arthur Street
Centerport, NY 11721
631-470-9446
www.apitherapy.org

A nonprofit membership organization devoted to advancing the investigation and promoting the use of honey bee products to further good health and to treat a variety of conditions and diseases.

American Beekeeping Federation (ABF)
P.O. Box 1337
115 Morning Glory Circle
Jesup, GA 31598-1038
912-427-4233
www.abfnet.org

"Serving the industry since 1943," the ABF promotes and assists in all aspects of beekeeping in the United States, from legislation to trade shows to research. This description just scratches the surface; there is much must-read information on their website.

International Bee Research Foundation (IBRA)
16 North Road
Cardiff CF1 3DY
United Kingdom
www.ibra.org.uk

This site has hundreds of links to scientific data on bees, apitherapy, recipes, beekeepers, and journals.

WEBSITES *of* INTEREST *to* BEEKEEPERS

There are many opportunities for learning more about bees and beekeeping on the Web, but these all have multiple links themselves and so are excellent starter or launch sites:

www.apitherapy.com

If you want to explore the world of apitherapy in more detail, this site offers tomes of research as well as discussion boards, where conferences are being held, even where to find practitioners of apitherapy.

www.badbeekeeping.com

This site describes itself as "A Thousand Great Places to Bee on the Web," and it literally has at least this many links to sites of interest all over the world. With everything from research to supplies to international beekeeping groups to bee-themed museums, this site is amazing.

www.bee-quick.com

Fischer's Bee Quick is a nontoxic, alcohol-free blend of natural oils and herbal extracts that cleans supers fast without foul odors.

www.care2.com
Care2 is concerned with community activism, green and healthy living, and more. Among its beeswax offerings: recipes for candles, lip balms, and a solvent-free furniture polish by "green" author Annie B. Bond.

www.exhibits.mannlib.cornell.edu/beekeeping/ack.html
More and more of the E. F. Phillips Beekeeping Library in the Mann Library at Cornell University is accessible online. It is a treat.

www.honey.com
The National Honey Board (NHB), a federal research and promotion board under USDA oversight, conducts research, marketing, and promotion programs to help maintain and expand domestic and foreign markets for honey. The NHB is not a regulatory agency nor does it have powers of enforcement. The ten-member board, appointed by the U.S. Secretary of Agriculture, represents producers (beekeepers), packers, importers, and a marketing cooperative.

www.honeybhealthy.com
Beekeepers Bob Noel and Jim Amerine developed Honey-B-Healthy, an earth-friendly product that controls tracheal mites and is safe to give to honeybees on a regular basis.

www.masterbeekeeper.org
The website for Cornell University's Dyce Laboratory for Honey bee Studies has a list of links to reports and findings on various aspects of bees and beekeeping.